Y0-BRR-925

Atoms in Chemistry: From Dalton's Predecessors to Complex Atoms and Beyond

ACS SYMPOSIUM SERIES **1044**

Atoms in Chemistry: From Dalton's Predecessors to Complex Atoms and Beyond

Carmen J. Giunta, Editor
Le Moyne College

Sponsored by the
ACS Division of the History of Chemistry

American Chemical Society, Washington, DC

541.24
A881

Library of Congress Cataloging-in-Publication Data

Atoms in chemistry: from Dalton's predecessors to complex atoms and beyond /
Carmen J. Giunta, editor.
    p. cm. -- (ACS symposium series ; 1044)
    Includes bibliographical references and index.
    ISBN 978-0-8412-2557-2 (alk. paper)
    1. Atomic theory--History--Congresses. I. Giunta, Carmen
    QD461.A863 2010
    541'.24--dc22
                                    2010023448

The paper used in this publication meets the minimum requirements of American National
Standard for Information Sciences—Permanence of Paper for Printed Library Materials,
ANSI Z39.48n1984.

PRINTED IN THE UNITED STATES OF AMERICA

Foreword

The ACS Symposium Series was first published in 1974 to provide a mechanism for publishing symposia quickly in book form. The purpose of the series is to publish timely, comprehensive books developed from the ACS sponsored symposia based on current scientific research. Occasionally, books are developed from symposia sponsored by other organizations when the topic is of keen interest to the chemistry audience.

Before agreeing to publish a book, the proposed table of contents is reviewed for appropriate and comprehensive coverage and for interest to the audience. Some papers may be excluded to better focus the book; others may be added to provide comprehensiveness. When appropriate, overview or introductory chapters are added. Drafts of chapters are peer-reviewed prior to final acceptance or rejection, and manuscripts are prepared in camera-ready format.

As a rule, only original research papers and original review papers are included in the volumes. Verbatim reproductions of previous published papers are not accepted.

ACS Books Department

Contents

Indexes

Chapter 1

Introduction

Carmen J. Giunta*

Department of Chemistry and Physics, Le Moyne College,
Syracuse, NY 13214
*giunta@lemoyne.edu

200 Years of Atoms in Chemistry: From Dalton's Atoms to Nanotechnology

This volume contains presentations from a symposium titled "200 Years of Atoms in Chemistry: From Dalton's Atoms to Nanotechnology," held at the 236th national meeting of ACS in Philadelphia in August 2008. The occasion was the 200th anniversary of the publication of John Dalton's *A New System of Chemical Philosophy* (*1*).

Dalton's theory of the atom is generally considered to be what made the atom a scientifically fruitful concept *in chemistry*. To be sure, by Dalton's time the atom had already had a two-millenium history as a philosophical idea, and corpuscular thought had long been viable in natural philosophy (that is, in what we would today call physics).

John Dalton (1766-1844) lived and worked most of his life in Manchester, and he was a mainstay of that city's Literary and Philosophical Society. He had a life-long interest in the earth's atmosphere. Indeed, it was this interest that led him to study gases, out of which study grew his atomic hypothesis (*2*). His experiments on gases also led to a result now known as Dalton's law of partial pressures (*3*). Dalton's name is also linked to color blindness, sometimes called daltonism, a condition he described from firsthand experience.

The laws of definite and multiple proportions are also associated with Dalton, for they can be explained by his atomic hypothesis. The law of definite proportions or of constant composition had previously been proposed in the work of Jeremias Richter and Joseph-Louis Proust. The law of multiple proportions came to be regarded as an empirical law quite independent of its relation to the atomic hypothesis or perhaps as an empirical law that inspired the atomic hypothesis; however, Roscoe and Harden have shown that in Dalton's mind it was a testable prediction which followed from the atomic hypothesis (*4*).

Dalton's 1808 *New System* (*1*) contains a detailed and mature presentation of his atomic theory. It is not, however, the first published statement of his atomic ideas or the first table of his atomic weights. A "Table of the relative weights of the ultimate particles of gaseous and other bodies" appears in reference (2), published in 1805 after having been read in 1803. Thomas Thomson's account of Dalton's theory (5) also preceded the publication of Dalton's book—with Dalton's permission.

Thus, 2008 was perhaps an arbitrary year to celebrate 200 years of Dalton's theory, but as good a year as any. The Symposium Series volume appears in 2010, which is 200 years after the publication of Part II of Dalton's *New System*. Readers interested in learning more about Dalton's life and work are directed to Arnold Thackray's 1972 volume which remains authoritative even after nearly four decades (6).

As originally envisioned, the symposium was to examine episodes in the evolution of the concept of the atom, particularly in chemistry, from Dalton's day to our own. Clearly, many of Dalton's beliefs about atoms are not shared by 21st-century scientists. For example, the existence of isotopes contradicts Dalton's statement that "the ultimate particles of all homogeneous bodies are perfectly alike in weight, figure, &c."(1) Other properties long attributed to atoms, such as indivisibility and permanence have also been discarded over the course of the intervening two centuries.

One property that remains in the current concept of atom is discreteness. If anything, evidence for the particulate nature of matter has continued to accumulate over that time, notwithstanding the fact that particles can display wavelike phenomena such as diffraction and regardless of their *ultimate* nature (quarks? multidimensional strings? something else?).

Images that resolve discrete atoms and molecules became available in the 1980s, with the invention of scanning tunnelling microscopy (STM). Its inventors, Gerd Binnig and Heinrich Rohrer, submitted their first paper on STM in fall 1981. Five years later, they were awarded the Nobel Prize in physics. Before long, other scientists at IBM turned an STM into a device that could pick up and place individual atoms, in effect turning atoms into individual "bricks" in nanofabricated structures.

STM was the first of a class of techniques known as scanning probe microscopy. Atomic force microscopy (AFM), invented later in the 1980s, is currently the most widely used of these techniques. Both STM and AFM depend on probes with atomically sharp tips; these probes are maneuvred over the surface of the sample to be imaged, maintaining atom-scale distances between the probe and sample. Both techniques are capable of picking up atoms individually and placing them precisely on surfaces (7).

Scanning probe microscopy and manipulation lie at the intersection of 21st-century nanotechnology and 19th-century Daltonian atomism. Never mind the fact that the devices depend on quantum mechanical forces: the devices also require atomic-scale engineering to make sharp tips and to steer the probes closely over sample surfaces. But more importantly, they make visible individual discrete atoms and are capable of manipulating them. As originally conceived, the symposium would have had a presentation on applications of atomism to nanotechnology to bring the coverage up to the present—or even the future. Alas, that presentation never materialized, but hints of what it might have covered remain in the introduction of this volume to give a sense of the sweep of the topic and its continued relevance to current science.

Atoms in Chemistry: From Dalton's Predecessors to Complex Atoms and Beyond

As already noted, the symposium did not include atoms in nanotechnology. Neither did it treat the quantum-mechanical atom. So the near end of the historical span actually included in the symposium extended to the first half of the 20th century. The far end of that span turned out to be closer to two millenia ago than two centuries. As a result, the title of the symposium series volume is *Atoms in Chemistry: From Dalton's Predecessors to Complex Atoms and Beyond*.

William B. Jensen begins the volume with an overview of scientific atomic theories from the 17th through 20th centuries. He mentions ancient atomism, but he begins in earnest analyzing corpuscular theories of matter proposed or entertained by natural philosophers in the 17th century. He describes the dominant flavors of atomic notions over four centuries, from the mechanical through the dynamical, gravimetric, and kinetic, to the electrical. Jensen is Oesper Professor of Chemical Education and History of Chemistry at the University of Cincinnati and was the founding editor of the *Bulletin for the History of Chemistry*.

Leopold May goes back even further in time to outline a variety of atomistic ideas from around the world. His chapter "Atomism before Dalton" concentrates on conceptions of matter that are more philosophical or religious than scientific, ranging from ancient Hindu, to classical Greek, to alchemical notions, before touching on a few concepts from the period of early modern science. May is Professor of Chemistry, Emeritus, at the Catholic University of America in Washington, DC.

The next two chapters jump to the middle of the 19th century, a time when many chemists were using atomic models while avowing a strict agnosticism about the physical nature or even physical reality of atoms.

David E. Lewis presents a sketch of 19th-century organic structural theories in a chapter entitled "150 Years of Organic Structures." Fifty years after Dalton, Friedrich August Kekulé and Archibald Scott Couper independently published representations of organic compounds that rationalized their chemisty and even facilitated the prediction of new compounds. The investigators did not assign any physical meaning to their structures, much less assert anything about the arrangement of atoms in space. Yet the models were inherently atomistic because they relied on the atomistic picture of bonding put forward by Dalton (that is, bonding atom to atom). Organic compounds behaved *as if* the carbon in them formed chains (*i.e.*, as if they were connected to each other atom to atom) and was tetravalent. Lewis is Professor of Chemistry at the University of Wisconsin-Eau Claire.

William H. Brock describes episodes from the second half of the 19th century in which chemists debated the truth of the atomic-molecular theory. In both cases, doubts about the physical reality of atoms led chemists to question the soundness of chemical atomism. The two central figures in this chapter are Benjamin Brodie, who proposed a non-atomic calculus of chemical operations in 1866, and Wilhelm Ostwald, who proposed to base chemistry on energetics in the 1890s. Brock is Professor Emeritus of History of Science at the University of Leicester in the United Kingdom. He is the author of numerous books and papers on the history of chemistry, including *The Norton History of Chemistry*.

The next two chapters turn to the physical evidence accumulated in the late 19th and early 20th centuries that suggested that atoms were actually real, even if they were not exactly as Dalton envisioned them.

The first of these chapters, by Carmen Giunta, concentrates on the evidence that atoms are composite—not the ultimate particles of matter. Evidence for the divisibility and impermanence of atoms was collected even while some chemists and physicists continued to doubt their very existence. The chapter focuses on discoveries of the electron, the nucleus, and the heavy particles of the nucleus. Giunta is Professor of Chemistry at Le Moyne College in Syracuse, New York, and he maintains the Classic Chemistry website.

The latter chapter, written by Gary Patterson, focuses on converging lines of evidence for the physical existence of atoms. By the early decades of the 20th century, through the efforts of Jean Perrin and others, skepticism over the physical existence of atoms was practically eliminated. Patterson describes evidence from X-rays, radioactivity, quantum theory, spectroscopy, and more—all converging on the physical existence of atoms and molecules. Gary Patterson is Professor of Chemistry and Chemical Engineering at Carnegie Mellon University in Pittsburgh, Pennsylvania.

The final chapter, by Jim and Jenny Marshall, takes the reader beyond the atom itself to some of the places associated with the history of scientific atomism. "Rediscovering Atoms: An Atomic Travelogue" takes the reader to several sites in Europe and North America where important work was done on the development of chemical atomism. The authors include photos of atom-related sites from their extensive DVD travelogue *Rediscovery of the Elements*. Jim Marshall is Professor of Chemistry at the University of North Texas in Denton, Texas, and Jenny Marshall is an independent contractor of computer services.

To physically visit the sites described by the Marshalls requires a passport. It is hoped that this volume itself can serve as a passport to important episodes from the more than 200-year history of atoms in chemistry.

References

1. Dalton, J. *A New System of Chemical Philosophy*; R. Bickerstaff: Manchester, U.K., 1808; Part I.
2. Dalton, J. On the absorption of gases by water. *Mem. Manch. Lit. Philos. Soc.* **1805**, *1*, 271–287.
3. Dalton, J. Experimental enquiry into the proportion of the several gases or elastic fluids, constituting the atmosphere. *Mem. Manch. Lit. Philos. Soc.* **1805**, *1*, 244–258.
4. Roscoe, H. E.; Harden, A. *A New View of the Origin of Dalton's Atomic Theory*; Macmillan: London, 1896.

5. Thomson, T. *A System of Chemistry*, 3rd ed.; Bell & Bradfute, E. Balfour: London, 1807; Vol. 3.
6. Thackray, A. *John Dalton: Critical Assessments of His Life and Science*; Harvard University Press: Cambridge, U.K., 1972.
7. Amato, I. Candid cameras for the nanoworld. *Science* **1997**, *276*, 1982–1985.

Chapter 2

Four Centuries of Atomic Theory

An Overview

William B. Jensen[*]

Department of Chemistry, University of Cincinnati, P.O. Box 210172,
Cincinnati, OH 45221
[*]jensenwb@ucmail.uc.edu

Introductory Apology

It might seem oddly perverse to give a lecture entitled "400 Years of Atomic Theory" at a symposium entitled "200 Years of Atoms in Chemistry." No one questions, of course, that the 19th- and 20th-centuries were the heyday of chemical atomism and historians of chemistry have long agreed that Dalton's work was the starting point for our current quantitative views on this subject. Less well known, however, is the fact that atomism had been slowly seeping into chemical thought for nearly two centuries before Dalton and, that while these earlier variants of chemical atomism did not lead to a significant breakthrough in chemical theory, they nonetheless gradually produced a significant qualitative reorientation in the way in which chemists thought about chemical composition and reactivity—a qualitative reorientation which formed an essential foundation for the rise of a quantified gravimetric atomism based on Dalton's concept of atomic weight.

My task in this overview lecture is to give you both a feel for this qualitative pre-Daltonian foundation and to properly interface this prehistory with the later developments of the 19th and 20th centuries, which will be the focus of the other talks in this symposium. I hope do this by presenting a very broad overview of how each century tended to focus on a different atomic parameter and how this changing focus was reflected in the chemical thought of the period.

Select Bibliography of Books Dealing with the General History of Atomism

- Brush, Stephen G. (1983), *Statistical Physics and the Atomic Theory of Matter from Boyle and Newton to Landau and Onsager*, Princeton University Press: Princeton, NJ.

- Gregory, Joshua (1931), A Short History of Atomism from Democritus to Bohr, Black: London.
- Kirchberger, Paul (1922), Die Entwicklung der Atomtheorie, Müllerische Hofbuchhandlung: Karlsruhe.
- Kubbinga, Henk (2003), De molecularisering van het wereldbeeld, 2 Vols.,Verloren: Hilversum.
- Lasswitz, Kurd (1890), Geschichte der Atomistik: Vom Mittelalter bis Newton, 2 Vols., Voss: Hamburg.
- Llosa de la, Pedro (2000), El espectro de Democrito: Atomismo, disidencia y libertad de pensar en los origines de la ciencia moderna, Ediciones del Serbal: Barcelona.
- Mabilleau, Leopold (1895), Histoire de la philosophie atomistique, Bailliére et Cie: Paris.
- Pullman, Bernard (1998), The Atom in the History of Human Thought, Oxford University Press: New York, NY.
- Pyle, Andrew (1995), Atomism and Its Critics: Problem Areas Associated with the Development of the Atomic Theory of Matter from Democritus to Newton, Thoemmes Press: Bristol.
- Van Melsen, Andreas (1952), From Atomos to Atom: The History of the Concept of Atom, Duquesne University: Pittsburg, PA, 1952.

Ancient Atomism

Before beginning our four-century survey, however, it is necessary to first say a little about ancient atomism—and by ancient atomism I mean the reductionistic mechanical atomism of Leucippus, Democritus and Epicurus rather than the nonreductionistic pseudo-corpuscularism associated with the "seeds" of Anaxagoras or the "natural minima" of Aristotle. Only secondary and often critical accounts of the atomic doctrines of Leucippus and Democritus have survived (e.g. in the writings of Aristotle), whereas four Epicurean documents have survived: three short letters on various topics reproduced by the 3rd-century AD writer Diogenes Laertius in his *Lives of Eminent Philosophers*, and a major Latin prose poem, *On the Nature of Things*, by the 1st century BC Roman author, Titus Lucretius Carus.

Epicurean atomism was predicated on five basic assumptions:

a. There is an absolute lower limit to particle divisibility—i.e., true minimal particles called "atoms" which are not only indivisible but also immutable and thus permanent.

b. There is an interparticle vacuum or void.

c. All interparticle interaction is due to collision and mechanical entanglement.

d. The only fundamental atomic properties are size, shape, and motion—all others are secondary psychological responses to various atomic complexes.

8

e.   There is no dichotomy between mind and matter, thus implying that the soul is both material and mortal.

Thus we see that Epicurean atomism was both materialistic and strongly reductionistic. Given that, within the broader context of Epicurean philosophy, this strong naturalistic tendency was also coupled with an overt attack on both religion and superstition, it comes as little surprise that Epicurean atomism was an anathema to early Christianity and that this philosophical school essentially disappeared after 500 AD. Indeed it is remarkable that anything managed to survive at all.

Though often applied to physical processes, such as weathering, evaporation and filtration, there are no examples of the application of ancient atomism to phenomena that we would today classify as chemical and hence our survey of its gradual modification and influence on chemistry does not truly begin until the 17th century.

Select Bibliography of Books Dealing with Ancient and Medieval Atomism

- Alfieri, Vittorio (1979), *Atomos idea: l'origine del concetto dell' atomo nel pensiero greco*, Galatina: Congedo.
- Bailey, Cyril (1928), *The Greek Atomists and Epicurus: A Study*, Clarendon: Oxford.
- Pines, Shlomo (1997), *Studies in Islamic Atomism*, Magnes Press: Jerusalem.

17th-Century Mechanical Atomism

While printed editions of both Diogenes Laertius and Lucretius were available by the 15th century—the first editions appearing in 1472 and 1473 respectively—it was not until the 17th century that atomism began to seriously impact on European science. A necessary prerequisite for this process was the "Christianization" of Epicurean atomism through elimination of its more objectionable assumptions, much as Thomas Aquinas and the scholastics had done four centuries earlier for the writings of Aristotle. This task was undertaken by the French priest and scientist, Pierre Gassendi, and by his English imitator, Walter Charleton, in the period 1640-1660. Atoms were no longer self-existent entities whose fortuitous collisions led to the creation of both the universe and man himself, but rather were instead created by God and directed by him for his own predetermined purposes. Boyle did much the same by the simple expedient of dissociating atomism from the despised names of both Epicurus and Lucretius and referring to it instead as either the "corpuscular doctrine" or the "Phoenician doctrine."

The revival of atomism in the 17th century is actually quite complex and involved not only the true mechanical atomism of Epicurus, but also various hybridized versions based largely on the reification and atomization of the older Aristotelian and Platonic theories of forms and seminal principles. Within

9

these hybridized versions, atoms could act as the inherent carriers of such secondary properties as color, taste, acidity, hotness and even coldness. These corpuscularized qualities would eventually evolve into the imponderable fluids much beloved of the 18th- and early 19th-century theorist, of which phlogiston and caloric are perhaps the best known examples.

In addition, several new forms of atomism or corpuscularism were also introduced, the most famous of which were Descartes' plenum theory and Newton's dynamic atomism, both of which rejected one or more of the basic assumptions of Epicurean atomism. Thus Descartes rejected both a lower limit to particle divisibility and the existence of an interparticle vacuum or void, as well as insisting on a strong dichotomy between matter and soul, whereas Newton replaced mechanical entanglement with short-range interparticle forces of attraction and repulsion.

It is well known that Robert Boyle was the major proponent of the application of particulate or corpuscular theories to chemical phenomena in the 17th-century, though neither he nor his contemporaries were able to develop a specific form of the theory which could be meaningfully related to quantitative chemical data. As a consequence, the true impact of mechanical corpuscularism on 17th-century chemistry was largely indirect and is best illustrated, as J. E. Marsh observed many years ago, in terms of its application to the acid-alkali theory of salt formation.

The reaction between various acids and various alkalis or metallic carbonates first attracted the attention of iatrochemical writers as a possible chemical model for the processes of digestion. Ignoring the carbon dioxide gas that was generated, which they misinterpreted as a violent churning or mechanical motion of the interacting particles, they viewed this reaction as a simple addition:

$$acid \ + \ alkali \ = \ salt$$

Acids were thought to have sharp, pointed particles, which accounted for their sour taste and ability to attack or corrode substances, whereas alkalis were thought to have porous particles. Neutralization and salt formation consisted in the points of the acid particles becoming mechanically wedged in the pores of the alkali particles, thus blunting or neutralizing their properties (Figure 1).

The importance of this theory for chemistry, however, did not lie in this mechanical mechanism for neutralization, but rather in the fact that it gradually accustomed chemists to the idea of characterizing salts in terms of their component acid and alkali particles rather than in terms of property-bearing principles and to looking at acid-alkali reactions as exchanges between preexisting material components, rather than in terms of the generation and corruption of alternative abstract forms or essences. This newer way of looking at neutralization reactions can be found in the writing of many 17th-century chemists, including Glauber, Lemery, Sylvius, Tachenius, and especially John Mayow, who would cite a laboratory example of the analysis and synthesis of various nitrate salts interpreted in terms of the separation and addition of their component acids and alkali particles.

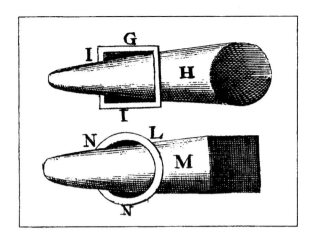

Figure 1. A typical 17th-century atomistic interpretation of acid-alkali neutralization in terms of points and pores. (From T. Craanen, Tractatus physico-medicus de homine, 1689).

Select Bibliography of Books Dealing with Seventeenth-Century Mechanical Atomism

- Boas, Marie (1958), *Robert Boyle and Seventeenth-Century Chemistry*, Cambridge University Press: Cambridge.
- Clericuzio, Antonio (2000), *Elements, Principles, and Corpuscles: A Study of Atomism and Chemistry in the Seventeenth Century*, Kluwer: Dordrecht.
- Kargon, Robert (1966), *Atomism in England from Hariot to Newton*, Clarendon: Oxford.

18th-Century Dynamical Atomism

As already noted, Newton replaced the concept of mechanical entanglement with the postulate of short-range interparticle forces of attraction and repulsion and applied this model in his *Principia* of 1687 to rationalize Boyle's law relating gas pressure and volume. However, it was not until the first decade of the 18th century that this new dynamic or force model was first specifically applied to chemical phenomena by the British chemists, John Freind and John Keill, and by Newton himself in the finalized version of the 31st query appended to the 1717 and later editions of his famous treatise on optics, where he succinctly summarized his new particulate program for chemistry:

There are therefore Agents in Nature able to make the Particles of Bodies stick together by strong Attractions. And it is the Business of experimental Philosophy to find them out.

11

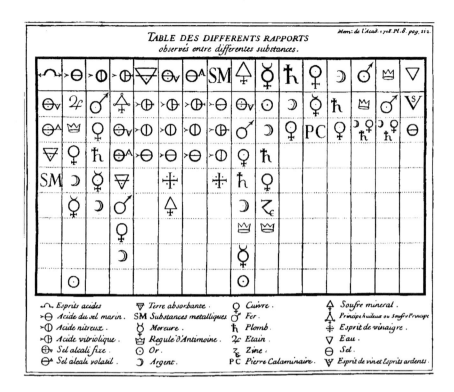

Figure 2. Geoffroy's 1718 affinity table for single displacement reactions interpreted as particle interchanges.

Meanwhile the particulate approach to chemical reactions, first realized in the 17th-century theory of acid-alkali neutralizations, was applied to chemical reactions in general, which were now being routinely classified as simple additions, simple decompositions, single displacements, and double displacements—an advance difficult to imagine within the older context of the theory of forms and essences which had dominated chemical thought for centuries. In addition, empirical observations concerning the observed outcomes of single displacement reactions were being tabulated, starting with the work of Geoffroy in 1718, in the form of so-called "affinity tables" (Figure 2), as well as in a series of textbook statements known as the "laws of chemical affinity" (e.g, Macquer 1749).

It was not long before this empirical concept of chemical affinity became associated with the concept of Newtonian short-range interparticle forces, an identification best expressed in Bergman's 1775 monograph, *A Dissertation on Elective Attractions*, and in attempts, now known to be flawed, by such chemists as Guyton de Morveau, Wenzel, and Kirwan to quantitatively measure these forces—attempts which also culminated in an early precursor of the chemical equation known as an "affinity diagram" (Figure 3).

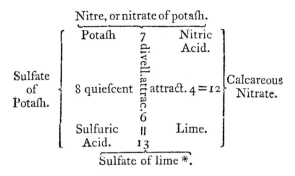

Figure 3. A typical late 18th-century affinity diagram. (From A. Fourcroy, Elements of Natural History and Chemistry, 1790).

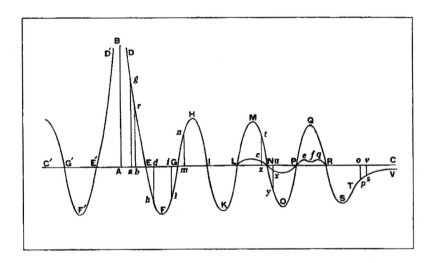

Figure 4. An 18th-century Newtonian force atom. (From R. Boscovitch, Theoria philosophiae naturalis, 1763).

As the concept of the Newtonian force atom came to dominate 18th-century chemical atomism, the parameter of atomic shape, so important to 17th-century mechanical atomism, faded and chemists and physicists came to more and more think of atoms as spherical—a view which reached its most extreme form in Roger Boscovitch's 1763 monograph, Theoria philosophiae naturalis, in which the atom was reduced to an abstract point for the convergence of a series of complex centro-symmetric force fields (Figure 4).

Select Bibliography of Books Dealing with Eighteenth-Century Dynamic Atomism

- Duncan, Alistair (1996), *Laws and Order in Eighteenth-Century Chemistry*, Clarendon: Oxford.
- Kim, Mi Gyung (2003), *Affinity that Elusive Dream*, MIT Press: Cambridge, MA.
- Thackray, Arnold (1970), *Atoms and Powers: An Essay on Newtonian Matter-Theory and the Development of Chemistry*, Harvard: Cambridge, MA.

19th-Century Gravimetric Atomism

This background now allows us to more fully appreciate the uniqueness of Dalton's contribution, when, in the period 1803-1808, he shifted, for the first time, the focus of chemical atomism from the atomic parameters of shape and interparticle forces to a consideration of relative atomic weights, with a concomitant emphasis on characterizing the chemical composition of individual species rather than on the classification and rationalization of chemical reactions.

By the end of the 18th century it was possible to characterize the chemical composition of a species at the molar level in terms of its composition by weight, or, in the case of gases, by its composition by volume. Thus one could speak of water as being composed of 11.11% hydrogen and 88.89% oxygen by weight or of 66.67% hydrogen and 33.33% oxygen by volume. With the introduction of the atomic weight concept, however, one could now characterize the composition of a species at the molecular level in terms of the relative number of component atoms and so speak of water as composed of molecules containing a ratio of two hydrogen atoms to one oxygen atom.

The key to Dalton's compositional revolution was the ability to link atomic weights at the molecular level with gravimetric composition measured at the molar level using his so-called "rules of simplicity." These, however, were soon shown to be operationally flawed and nearly a half century would pass before this problem was finally solved in a satisfactory manner by Cannizzaro in 1858 and accepted by the chemical community at the Karlsruhe conference of 1860. This final resolution of the problem of chemical composition was, of course, soon brilliantly elaborated by the rise of chemical structure theory and classical stereochemistry during the last quarter of the 19th century. The story of these advances is, of course, far more complex and nuanced then suggested by this brief summary and aspects of it will no doubt be covered in greater detail by other speakers in this symposium.

Select Bibliography of Books Dealing with Nineteenth-Century Gravimetric Atomism

- Bradley, John (1992), *Before and After Cannizzaro: A Philosophical Commentary on the Development of the Atomic and Molecular Theories*, Whittles Publishing: Caithness, UK.
- Brock, William, Ed. (1967), *The Atomic Debates: Brodie and the Rejection of the Atomic Theory*, Leicester University Press: Leicester.
- Meldrum, Andrew (1904), *Avogadro and Dalton: The Standing in Chemistry of their Hypotheses*, Clay: Edinburgh.
- Mellor, D. P. (1971), *The Evolution of the Atomic Theory*, Elsevier: Amsterdam.
- Rocke, Alan (1984), *Chemical Atomism in the Nineteenth Cemtury: From Dalton to Cannizzaro*, Ohio State University Press: Columbus, OH.

19th-Century Kinetic Atomism

If the gravimetric Daltonian atom was the chemist's primary contribution to atomic theory in the 19th century, then the kinetic atom was the physicist's primary contribution. Atomic motion was, of course, always a part of the atomic theory from ancient atomism onward. However, it functioned primarily as a way of explaining diffusion and providing a means for bringing about sufficient contact between particles to facilitate either mechanical entanglement or the engagement of short-range forces of attraction and repulsion. Aside from this minimal function, motion played little role in explaining the properties of things in either 17th-century mechanical atomism or in 18th-century dynamical atomism.

Thus, within the context of the Newtonian force atom and the caloric theory of heat, solids, liquids, and gases were all viewed as organized arrays of particles produced by a static equilibrium between the attractive interparticle forces, on the one hand, and the repulsive intercaloric forces, on the other. The sole difference was that the position of equilibrium became greater as one passed from the solid to the liquid to the gas, due to the increasing size of the caloric envelopes surrounding the component atoms (Figures 5 and 6).

Likewise, Berthollet's original concept of chemical equilibrium, introduced in the years 1799-1803, was also based on the concept of a static equilibrium between those forces favoring the formation of the products versus those favoring the formation of the reactants. As is well known, this static model made it very difficult to rationalize the law of mass action without coming into conflict with the law of definite composition. This static view of both states of matter and chemical equilibrium, viewed as a competition between chemical affinity and caloric repulsions, continued to dominate chemical thought throughout the first half of the 19th century.

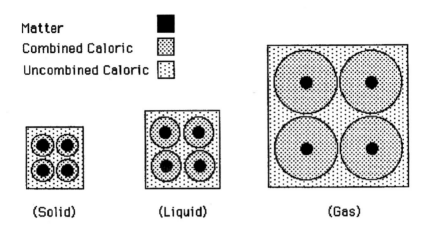

Matter

Combined Caloric

Uncombined Caloric

(Solid)   (Liquid)   (Gas)

Figure 5. The author's graphical interpretation of the caloric theory of states.

Figure 6. Daltonian atoms and molecules with their surrounding atmospheres of repulsive caloric. (From J. Dalton, A New System of Chemical Philosophy, Part. II, 1810).

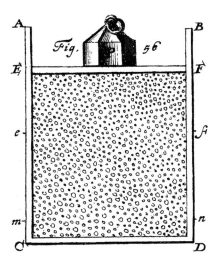

Figure 7. The first known attempt to envision gas pressure in terms of a kinetic model of atoms and molecules. (From D. Bernoulli, Hydrodynamica, 1738).

Though a kinetic model of gases had been proposed by Bernoulli as early as 1738 (Figure 7) and was unsuccessfully revived by Herapath (1821) and Waterson (1845) in the first half of the 19th century, it was not until the 1850s and 1860s that it began to attract widespread acceptance through the work of Krönig (1856) and Clausius (1857) in Germany and Joule (1848) and Maxwell (1859) in England. Heat was no longer a self-repulsive imponderable fluid but rather a measure of the average kinetic energy of molecular motions. States of matter were no longer the result of a static equilibrium between attractive interparticle forces and repulsive intercaloric forces, but rather the result of a dynamic equilibrium between attractive interparticle forces and disruptive thermal motions. Solids, liquids and gases no longer shared a common structure, differing only in their distance of intermolecular equilibration, but now differed in terms of both their degree of intermolecular organization and their freedom of motion. Chemical equilibrium and mass action were no longer a static equalization of opposing forces, but rather a dynamic equilibrium based on relative collision frequencies and differing threshold energies for reaction—a view first qualitatively outlined by the Austrian physicist, Leopold Pfaundler, in 1867.

Thus by 1895, the German chemist, Lothar Meyer, would conclude the short version of his textbook of theoretical chemistry with the observation that:

Chemical theories grow more and more kinetic.

a trend which would culminate in the development of classical statistical mechanics by Boltzmann and Gibbs by the turn of the century and which would continue unabated throughout the 20th century.

17

Select Bibliography of Books Dealing with Nineteenth-Century Kinetic Atomism

- Brush, Stephen (2003), *The Kinetic Theory of Gases: An Anthology of Classic Papers with Historical Commentary*, Imperial College Press: River Edge, NJ.
- Brush, Stephen (1986), *The Kind of Motion We Call Heat: A History of the Kinetic Theory of Gases in the 19th Century*, 2 Vols., North Holland; Amsterdam.

20th-Century Electrical Atomism

With the advent of the 20th-century we see the solid, impenetrable, billard-ball atom of the previous centuries replaced by the diffuse, quantized electrical atom (Figures 8 and 9). Nevertheless the various atomic parameters emphasized by earlier variants of atomism have all retained their importance in one way or another:

Like 17th-century mechanical atomism, modern atomism also recognizes the importance of shape—at the level of individual atoms in terms of the concept of orbital hybridization and directional bonding—and at the molecular level in terms of the lock and key model of intermolecular interactions.

Like 18th-century dynamical atomism, modern atomism also recognizes the importance of short-range interparticle forces—now interpreted in terms of electrical forces of attraction and repulsion between negatively charged electrons and positively charged nuclei.

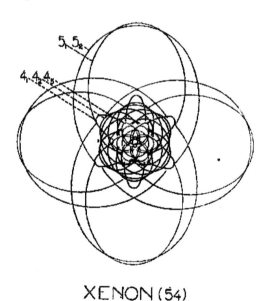

XENON (54)

Figure 8. A Bohr-Sommerfeld model of the xenon atom. (From H. A. Kramers and H. Horst, The Atom and the Bohr Theory of its Structure, 1924).

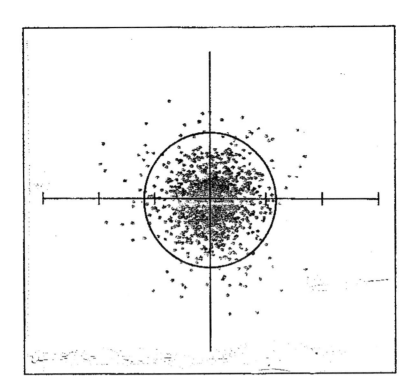

Figure 9. A modern statistical picture of the electron cloud of a hydrogen atom.

Like 19th-century gravimetric atomism, the concept of atomic weight and the laws of stoichiometry are still the cornerstones of chemical composition—albeit now modified to accommodate the concepts of isotopes and relativistic mass effects.

Like 19th-century kinetic atomism, molecular motion still forms the cornerstone of our modern understanding of heat, thermodynamics, kinetics, and statistical mechanics, but now also plays a key role in our understanding of the internal structure of the atom itself, via the concept of quantized electron motions.

Select Bibliography of Books Dealing with Twentieth-Century Electrical Atomism

- Hund, Friedrich (1974), *The History of Quantum Mechanics*, Harper & Row: New York, NY.
- Keller, Alex (1983), *The Infancy of Atomic Physics: Hercules in his Cradle*, Clarendon: Oxford.
- Stranges, Anthony (1982), *Electrons and Valence: Development of the Theory, 1900-1925*, Texas A & M: College Station, TX.

Chapter 3

Atomism before Dalton

Leopold May*

Department of Chemistry, The Catholic University of America,
Washington, DC 20064
*may@cua.edu

In ancient times, the notion of small particles or atoms making
up matter was conceived by philosophers first in Hindu India and
then in the Mediterranean (Greek) region. Kanada developed
an atomic theory in India where five elements were known,
air, water, fire, earth, and space. In the Mediterranean (Greek)
region, Democritus and Leucippus are considered to be the
founders of the *atomism* in the fifth century BCE. Aristotle did
not accept the atomic theory but did accept the four elements,
air, water, earth, and fire. Among the Arab alchemists, there was
little interest except for followers of Kalàm who developed an
atomic theory. In Europe, the Aristotelian view dominated until
the sixteenth century CE. The four elements of Empedocles
(earth, air, water, fire) or the three principles of Paracelsus
(mercury, sulfur, salt) were not included in Lavoisier's Table of
Simple Substances in 1789 CE. In the eighteenth century, there
was the revival of the ancient Greek atomism in the guise of
corpuscularism preceding the atomism of John Dalton.

The notion of atoms arrived in the East, ancient India, prior to its appearance
in the West, the ancient Mediterranean (Greek) world. Both societies were
polytheistic, and philosopher-chemists dominated the study of chemistry. Atomic
concepts were based upon philosophical considerations and not experimental
observations. No exchange on atomism between these two regions in this
ancient time has been detected, indicating that these concepts were developed
independently (*1*). These developments occurred during Period I of the Ancient
Regime of Chemistry (~10,000 BCE - ~100 BCE), which may be called the

era of Philosophical Chemistry because the philosophers of the time were the theoretical chemists. Of course, there were other forms of chemistry that were practiced. For example, the first recorded industrial chemists were two women, Tapputi-Belatekallim, the Perfumeress and ...Ninu, the Perfumeress, in Babylonia in about 2000 BCE. The full name of ...Ninu is not known due to lacunae in a cuneiform tablet, which was written in Akkadian, the language of Mesopotamia during the reign of Tukulti-Ninurti I (1256-1209 BCE) (2).

Indian Atomism

The atomic concept developed differently in the various religions prevalent in ancient India. These views survived until after the British conquest in the 18th century when the educational system was revamped to emulate the British educational system.

Hinduism

In the *Bhagavad Gīta*, one of the holy books of Hindus, which was written between 300 and 500 BCE (3), a reference to atoms appears in verse 9, chapter 8. It is written in Sanskrit: *kavim purāṇam anuśāsitāram aṇor aṇīyāmsam anusmared yah sarvasya dhātāram acintya-rūpam āditya-varṇam tamasah parastāt*, where *aṇor* refers to atom. One translation is (4): He who meditates on the one who is all-perceiving, the ancient, the ruler of all things, *smaller than the atom*, the supporter of this universe, whose form is inconceivable, who is as radiant as the sun beyond the darkness. Swami Prabhupada (5) offfers a different translation: "One should meditate upon the Supreme Person as the one who knows everything, as He who is the oldest, who is the controller, *who is smaller than the smallest*, who is the maintainer of everything, who is beyond all material conception, who is inconceivable, and who is always a person. He is luminous like the sun and, being transcendental, is beyond this material nature." In his commentary on this verse (5), he states "He is called the smaller than the smallest. As the Supreme, He can enter into the atom."

Kanada, a Nyaya-Vaisheshika philosopher, who lived ~600 BCE, considered that matter was composed of four types of atoms, earth, fire, air, and water. Atoms reacted with the aid of an invisible force (*adrsta*) to form biatomic molecules and triatomic molecules (6–8). He stated that there were five elements: earth, fire, air, water, and space. Each atom also had qualities such as odor, taste, color and a sense of touch (8).

Jainism

In Jaina atomism (~900 BCE), the atom was the indivisible particle of matter. Each atom had attributes such as color, taste, and odor, as well as tactile qualities such as roughness or moistness. Atoms existed in space. The combination of atoms was produced by the differences in attributes such as roughness (8).

Buddhism

In the Sarvāstivādin school of Buddhism (~400 BCE), the minimum indivisible particle of matter was called the atom, which expresses the nature of matter. The characteristic atoms were earth (solid), water (liquid), fire (heat), air (moving), color, taste, odor, and sense of touch, and they existed in space. The smallest composite unit was considered to be composed of seven characteristic atoms, which are set at the apices and center of octahedron (*8*).

Chinese Atomism

Based upon Taoist philosophy, alchemy in China developed. Although there is not any literature concerning atomism among the ancient Chinese alchemists, five elements (Wu Xing) were acknowledged in the twelfth century BCE. These elements were water, fire, wood, gold or metal, and earth. The elements were frequently associated or matched with other sets of five, such as virtues, tastes, colors, tones, and the like (*9*). In about 1910, modern atomism probably came to China when Sun Yat Sen introduced modern European education.

Mediterranean (Greek) Atomism

On the other side of the Ancient world, thinking about atoms was initiated by Sanchuniathon of Sidon in Phoenicia around 1200 BCE (*10*). As first principles, he considered air and ether. Poseidonios (135-51 BCE) stated that Sanchuniathon "originated the ancient opinion about atoms" according to Strabo, a geographer and writer in the ancient world (*10*). Robert Boyle in the seventh century, CE, noted that Mochus or Moschus of Sidon was the first to devise an "atomical hypothesis" (*11*). This Moschus should not be confused with the poet of the same name of Syracuse (*12*) nor the philosopher of the same name of Elis (*13*), both of whom lived at a later period.

Four Elements

North of Phoenicia on the west coast of Asia Minor in the city of Miletus, Thales (630-550 BCE), the first Greek philosopher, taught that the primary substance was water on which the earth floats, and all things contain gods (*14*). His pupil, Anaximander or Aleximandros (611-545 BCE) replaced water with *apeiron* (*15*). The primary substance according to Anaximenes (585-525 BCE), who succeeded Anaximander, was air or breath. By condensation, it became wind, cloud, water, earth, and stone and by rarefaction, fire (*16*). Fire was the choice of Heraclitus (540-450 BCE) of Ephesus (Asia Minor) as the primary substance (*17*). Xenophanes (550-450 BCE) of nearby Colophon suggested that earth was the primary substance (*18*).

On the island of Sicily in the city of Akragos (Agrigentum), Empedocles (483-430 BCE) proposed a theory of four primordial substance or roots. He associated them with deities, the identity of which varied with the source; Zeus (air or fire), Hera (air or earth), Aidoneus (air, earth, or fire), and Nestis (water) (*19*). Each root consisted of particles that were indivisible, homogeneous, changeless, and eternal with pores (not void) between the particles. The particles move with Love as the physical agent for mixing of the particles and for their separation, Strife (*19*). Until the end of the eigtheenth century, CE, this theory of four elements (seeds) persisted with the addition of mercury, sulphur, and salt.

First Defined Atomism

In contrast to Eleatic School (Parmenides and Empedocles), Leucippus of Miletus (~500-? BCE) and his pupil, Democritus of Abdera (460-370 BCE) introduced the void as being necessary for the motion of corpuscules or atoms. Atoms are indivisible, solid, full, and compact with various shapes. They also were in motion and have weight (*20*).

Born in Athens, using Pythagorarian concepts, Plato (427-347 BCE), a pupil of Socrates, conceived geometric bodies for the units or particles of the seeds, which he called elements. Earth units were cubes, fire units, tetrahedrons, air units, octahedrons, and water units, icosahedrons (*21*). He did not accept the void but thought that space existed inside the units. The units of fire, air, and water were deformable corpuscles. In his dialogue, *Timaeus*, he wrote "God placed water and air in the mean between fire and earth, and made them to have the same proportion so far as was possible (as fire is to air so is air to water, and as air is to water so is water to earth); and thus he bound and put together a visible and tangible heaven. And for these reasons, and out of such elements, which are in number four, the body of the world was created" (*22*).

Aristotle of Stageiros (384-322 BCE) did not agree with his teacher's geometric bodies for the different elements. He rejected the Democritian atoms in which matter was considered a principle but form was a secondary characteristic. Nor did he accept the existence of a void. According to the Aristotelian view, the four elements arose from the action on primordial matter by pairs of qualities (warm + dry, fire, warm + moist, air, cold + dry, earth, cold + moist, water). He introduced another element, ether, as a divine substance of which the heavens and stars are made (*23*).

Lucretius (Titus Lucretius Carus, ~99 BCE - ~55 CE) of Rome wrote a poem, *De Rerum Natura* (On the Nature of Things) (*24*) in which he described the atomic theory of Epicurus of Samos (342-271 BCE). For Epicurus, atoms were indivisible, invisible, and indestructible, and they differ in size, shape and weight. He believed that a void exists because there can be no motion of the atoms without it. The motions of atoms included the downward motion of free atoms because of their weight, "swerve," the deviation of atomic motion from straight downward paths, and "blow," which results from collisions and motion in compound bodies. Lucretius called atoms poppy seeds, bodies, principals, and shapes (*25*).

Galen of Pergamum (129-216 CE) rejected the atomic theory because the grouping of atoms could not explain why the properties of a compound differed from the properties of its constituents (*26*). His rejection effectively exiled atomism in the Western world in which the views of Aristotle prevailed until the seventeenth century, CE (*27*).

In 1615 CE, Cardinal Robert Bellarmine in his book, *The Mind's Ascent to God*, searched nature for lessons for the soul. No mention of atomism appeared. Sixty-five years later, Ralph Cudworth summarized a hypothesis of the time called atomical, corpuscular, or mechanical in his book, *The True Intellectual System of the Universe (28)*. Atomism had returned. How did this happen?

Alchemy

The second period of the Ancient Regime of Chemistry, alchemy, alchemi, alchimi or chymistry, began ~100 BCE and continued until the end of the eighteenth century, CE. In Egypt, the priests engaged in secret alchemical operations. As a result of this association of alchemy with the priests, alchemy became identified with magic. After Rome conquered Egypt and Emperor Constantine converted to Christianity, the administration of the empire was dominated by Christians intolerant of those who did not agree with the official views. Many of the alchemists were Gnostics exiled from the Roman Empire in the fifth century, CE. Also expelled were the Nestorians who carried the writings of the Greek philosophers, which were translated into Syrian in Persia. After Mohammad's death in 632 CE, his followers from Arabia created an empire from Persia to Spain. In Persia, the Greek texts including alchemical tracts were translated into Arabic *(29)*.

Arabian Atomism

There was little interest in atomism except for the followers of the philosophy of Kalàm (Arabic: speech). Among the main proponents were the *Mutazilites* (from i'tazala, to separate oneself, to dissent). Of the twelve propositions of Kalàm, the first nine were directly related to atoms. These propositions include:
"All things are composed of atoms that are indivisible, and when atoms combine, they form bodies."
"There is a vacuum."
"Time is composed of time-atoms"
"Substances cannot exist without accidents". Accidents are properties such as color, taste, motion or rest, and combination or separation.
"Atoms are furnished with accidents and cannot exist without them"
"Accidents do not continue in existence during two time-atoms. God creates substances and the accidents" *(30)*.

Medieval European Alchemy

In the twelfth and thirteenth centuries in Europe, Greek and Arab texts were translated from Arabic into Latin, the literary language of Europe. The first translation of an alchemical book from Arabic, *The Book of the Composition of Alchemy*, was prepared by Robert of Chester in 1144 CE in Spain *(31)*. To the Four Elements, air, water, fire, and earth, Arab alchemists added mercury and sulfur. Paracelsus considered mercury and sulfur as principles along with salt

(*32*). Aristotelian atomism was the only view accepted by the alchemists and the authorities (Catholic Church). Few if any references to atoms were made by alchemists in their writings. For example, in *The Ordinall of Alchemy* (1477) by Thomas Norton of Bristoll, atoms are mentioned once in the 123 pages (*33*):

Substance resolving in Attomes with wonder

Sympathizers and Atomists of the Twelfth to the Fourteenth Centuries, CE

The sympathizers, Adelard of Bath (died ~1150 CE) and Thierry of Chartres (died ~1155 CE), accepted the four elements of the Greeks and that atoms or corpuscles were involved with them. Atoms were fundamental constituents of substances according to Constantine the African of Carthage (twelfth century, CE). William of Conches (1080-1154 CE) recognized God's action as giving rise to the laws of nature and regarded atoms as "first principles" and "simple and extremely small particles" (*34*). Critizing Aristotelian physics, William of Ockham (1399-1350 CE) stated that substance had matter and form; its qualities result from elementary particles that can be construed to be atoms. In 1340 CE, his views were condemned by the Church as was those of Nicholas of Autrecourt (1300-1350 CE) in 1347 CE. Nicholas considered matter to be eternal, consisting of invisible atoms that are in motion; generation and corruption of substances occurs by the rearrangement of atoms (*34*).

In the early fifteenth century (1417 CE), *De Rerum Natura* by Lucretius was rediscovered. It was printed fifty-six years later in 1473 CE reintroducing the Epicurean concept of the atom and void to the western world (*35*).

Atomists of the Sixteenth to the Eighteenth Centuries, CE

One of the first atomists in the sixteenth century was Jean Bodin (1530-1596 CE) who considered atoms to be indivisible bodies with motion and that an infinite force was necessary for the division of atoms (*36*).

Giordano Bruno (1548-1600 CE) was a member of the Dominican order. His views on atoms had both metaphysical and physical aspects: atoms are both the ultimate, indeterminate, substance of things and a hypothesis that can be used to explain variety in the material world (even though only earth, among the four elements, has atoms). Each kind of being had a "minimum" or unit, although only God is a true monad; the point was the minimum of space, the atom the minimum of matter. Bruno's atoms are spherical, and their motions due to a soul in each. He was burnt alive for heresy on February 17, 1600, in Rome (*37*).

A professor of medicine at Wittenberg, Daniel Sennert (1572-1637 CE) developed a version of atomism from experimental observations rather than philosophical considerations. Based upon sublimation, solution, and petrifaction, for example, the mixtures of gold-silver alloy and silver dissolved in acid, he concluded that there were corpuscles or "minima" that were divisible, and the four elements had them (*36, 38*).

The atomism of Sebastian Basso (17th century, CE), a French physician, was based upon Democritus atomism with no void. He considered all bodies created by God to consist of exceedingly small atoms of different natures with the spaces between particles filled with a subtle spirit. The element fire consisted of fine and sharp corpuscles (*37, 39*).

Not an ardent supporter of Democritian atomism, Sir Francis Bacon, first Viscount St. Alban, (1561-1626 CE) was also a lawyer and member of the English government. He considered atoms to be true or useful for demonstration but he did not accept the void. The properties of bodies were explained by the size and shape of corpuscles and not the indivisible atoms. Force or motion was implanted by God in the first particles (*40*).

Isaac Beeckman (1588-1637 CE), a Dutch natural philosopher, proposed a "molecular" theory in his "scientific diary". He assumed that there were four kinds of atoms corresponding to the four elements of the one sole primordial matter. He considered these atoms to be the cause of the properties of the substances, for example, color, taste, smell, etc. The molecules of substances were called *homogenea physica* (physical homogenea) and were composed of the atoms in specific spatial structure. His private diary was available to several savants such as Descartes, who acknowledged these ideas in several books (*41*).

Another atomist, prosecuted by the Italian church authorities, was Galileo Galilei (1564-1642 CE). He initially used *minimi* to describe the smallest parts of substances but later applied the term to Epicurean atoms separated by a quantitatively infinite vacuum. The atomic structure of substances was necessary from mathematical reasoning, and the atom was indivisible without shape and dimensions. The qualities or properties (color, odor, taste, etc.) of atoms were not associated with atoms but with their sensory detection by the observer (*42*).

Two French contemporary students of the atomic theory were Pierre Gassendi (1592-1655 CE) and René Descartes (1596-1650 CE). Gassendi, a priest, was an atomist for whom atoms were primordial, impenetrable, simple, unchangeable, and indestructible bodies with shape, size, and weight that were set in motion by God at creation. In addition, he accepted that a vacuum exists, which Torricelli demonstrated in 1643 (*43*). In contrast, Descartes did not believe in the void, but that the material universe consisted of one infinite and continuous extended matter created by God. Extended matter consisted of a granulated continuum made of corpuscles. This corpuscular philosophy involved corpuscles that were deformable and divisible, having shapes, sizes, and motion (*44*). The association of God with atoms (or at least corpuscles) by Descartes and Gassendi was very instrumental in the return of Epicurean atomism as the basis of the atomic theory, and in 1678, Cudworth could include atomism in his book, *The True Intellectual System of the Universe* (*28*). Atomism had returned.

Nicholas Lemery (1645-1715 CE) was a corpuscularian who favored a five-element theory (water, spirit, oil, salt, and earth). His acid/alkali theory invoked spikes on an acid that interacted with the pores of the base. In 1675 CE in Paris, he published *Cours de Chymie*, a textbook that was translated into English, German, Italian, Latin, and Spanish and was popular for more than fifty years. In this book, he espoused the Cartesian corpuscular mechanism (*45*).

The corpuscularism of the Honorable Robert Boyle (1627-1691 CE) was based upon the theories of Descartes and Gassendi. He considered that matter was composed of corpuscles of different shapes, sizes, motion or rest, and solidity or impenetrability that are created by God. The four elements of Empedocles (earth, air, water, fire) or the three principles of Paracelsus (mercury, sulfur, salt) were not regarded as elements by him because he did not consider any of these to be fundamental constituents of existing bodies. However, he did not describe any elements. He exerted an extremely important influence on the development of chemistry as a science in the seventeenth century (46).

The prominent physicist, Sir Isaac Newton (1642-1727 CE), was also an alchemist. His theory of matter was in agreement with the atomism of Epicurus and Boyle including the existence of the void. In Query 31, Book 3, of *Opticks*, he wrote "All these things being consider'd, it seems probable to me, that God in the Beginning form'd Matter in solid, massy, hard, impenetrable Particles, of such Sizes and Figures, and with such other Properties, and in such Proportion to Space, as most conduced to the End for which he form'd them" (47). God's continual presence was also necessary for their continued existence. Newton assumed that the forces in corpuscles were not only gravitational but also had electrical, magnetic, attractive, and repulsive components (48).

Ruggiero Giuseppe Boscovich (1711-1787 CE), a Jesuit priest, replaced corpuscles with force-atoms (1758 CE) or point-centers of alternating attractive and repulsive forces. The views of Father Boscovich were similar to those of Newton, the Hindu atomists of the Nyaya-Vaisheshika, and the Arab followers of the Kalam (49).

The Russian atomist, Mikhail Vasilyevich Lomonosov (1711-1765 CE) believed that changes of matter were due to the motions of constituent particles. The particles consisted of "elementa" that contain no smaller bodies of different kinds. If a corpuscle (a small mass consisting of aggregates of elementa) consisted of the same elementa, it was homogeneous. If the components of the corpuscles were different elementa, the corpuscles were heterogeneous (50).

Bryan Higgins (1737 or 1741-1818) applied Newton's repulsion of atoms in air to simple and compound gases, and suggested that there were caloric atmospheres around molecules of compound gases (51). Many of his ideas were promoted by his nephew, William Higgins (1762/3-1825), who anticipated parts of Dalton's atomic theory and law of multiple proportions in 1789 (52). In 1814, he wrote (53):

These considerations gave birth to that doctrine which Mr. Dalton, eighteen years after I had written, claimed as originating from his own inventive genius. What his pretensions are will be seen from the sketches which will soon follow, and which have been taken from my book.

A controversy ensued concerning the awarding of credit with Dalton being remembered rather than Higgins (54).

In 1789, Antoine-Laurent Lavoisier (1743-1794 CE) published a *Table of Simple Substances* (p. 175-176) in his book *Traité élémentaire de Chimie, presenté dans un ordre nouveau at d'apres les découvertes modernes*, (55). The subtitle to the table was "Simple substances belonging to all the kingdom of nature, which may be considered as the elements of bodies". None of the four elements of Empedocles (earth, air, water, fire) or the three principles of Paracelsus (mercury, sulfur, salt) were included except for caloric one of whose old names was fire. On page xxiv, he wrote that "if by the term *elements*, we mean to express those simple and indivisible atoms of which matter is composed, it is extremely probable we know nothing at all about them." Thus, the Ancient Regime of Alchemy was overthrown, and the science of chemistry replaced it.

In ancient India and Greek lands, a notion of atoms
Philosophical chemists were making
After Aristotle, they were ceasing
Alchemy or the ancient régime II was beginning
Little were alchemists adding
Until chemistry, chymistry replacing
The atoms of Dalton were besting

References

1. Stillman, J. M. *The Story of Alchemy and Early Chemistry*; Dover Publications, Inc.: New York, 1960; p 105.
2. Levey, M. In *Great Chemists*; Farber, E., Ed.; Interscience: New York, 1961; p 3.
3. Jinarajadasa, C. *The Bhagavad Gita*; Theosophical Publishing House: Adyar, Madras, India, 1915; http://www.theosophical.ca/adyar_pamphlets/AdyarPamphlet_No59.pdf (accessed February 2, 2010).
4. *The Bhagavadgītā*; Nabar, V., Tumkur, S., Translators; Wordsworth Editions: Ware, U.K., 1997; p 36.
5. Swami Prabhupada, A. C. B. *Bhagavad-gītā As It Is*; Collier Books: New York, 1972; p 419.
6. Stillman, J. M. *The Story of Alchemy and Early Chemistry*; Dover Publications, Inc.: New York, 1960; pp 109−111.
7. (a) Horne, R. A. *Ambix* **1960**, *8*, 98−110. (b) Kanada. Wikipedia. http://en.wikipedia.org/wiki/Kanada (accessed February 2, 2010).
8. (a) Ohami, I. *Jap. Stud. Hist. Sci.* **1967**, *6*, 41−46. (b) Pullman, B. *The Atom in the History of Human Thought*; Reisinger, A., Translator; Oxford University Press: New York, 1998; Chapter 7.
9. Read, J. *Prelude to Chemistry*; The MIT Press: Cambridge, MA, 1939; p 20.
10. Partington, J. R. *A History of Chemistry*; Macmillan & Co., Ltd.: London, 1970; Vol. 1, pp 3−4.
11. Gregory, J. C. *A Short History of Atomism*; A. & C. Black, Ltd.: London, 1931; p 1.
12. Moschus. *Encyclopedia Britannica*, 11 ed.; The Encyclopedia Britannica Co.: New York, 1911; Vol. 18, p 891b.
13. Phaedo. *Encyclopedia Britannica*, 11 ed.; The Encyclopedia Britannica Co.: New York, 1911; Vol. 21, p 341b.
14. (a) Partington, J. R. *A History of Chemistry*; Macmillan & Co., Ltd.: London, 1970; Vol. 1, pp 6−7. (b) Pullman, B. *The Atom in the History of Human*

Thought; Reisinger, A., Translator; Oxford University Press: New York, 1998; pp 13−16.

15. (a) Partington, J. R. *A History of Chemistry*; Macmillan & Co., Ltd.: London, 1970; Vol. 1, pp 7−8. (b) Pullman, B. *The Atom in the History of Human Thought*; Reisinger, A., Translator; Oxford University Press: New York, 1998; pp 16−17.

16. (a) Partington, J. R. *A History of Chemistry*; Macmillan & Co., Ltd.: London, 1970; Vol. 1, p 8. (b) Pullman, B. *The Atom in the History of Human Thought*; Reisinger, A., Translator; Oxford University Press: New York, 1998; pp. 17−18.

17. (a) Partington, J. R. *A History of Chemistry*; Macmillan & Co., Ltd.: London, 1970; Vol. 1, pp 8−11. (b) Pullman, B. *The Atom in the History of Human Thought*; Reisinger, A., Translator; Oxford University Press: New York, 1998; pp 18−19.

18. (a) Partington, J. R. *A History of Chemistry*; Macmillan & Co., Ltd.: London, 1970; Vol. 1, pp15−17. (b) Pullman, B. *The Atom in the History of Human Thought*; Reisinger, A., Translator; Oxford University Press: New York, 1998; pp 19−20.

19. (a) Partington, J. R. *A History of Chemistry*; Macmillan & Co., Ltd.: London, 1970; Vol. 1, pp 17−23. (b) Pullman, B. *The Atom in the History of Human Thought*; Reisinger, A., Translator; Oxford University Press: New York, 1998; pp 21−24.

20. (a) Gregory, J. C. *The Atom in the History of Human Thought*; Reisinger, A., Translator; Oxford University Press: New York, 1998; pp 1−4. (b) Partington, J. R. *A History of Chemistry*; Macmillan & Co., Ltd.: London, 1970; Vol. 1, pp 35−49. (c) Pullman, B. *The Atom in the History of Human Thought*; Reisinger, A., Translator; Oxford University Press: New York, 1998; pp 31−37.

21. (a) Gregory, J. C. *The Atom in the History of Human Thought*; Reisinger, A., Translator; Oxford University Press: New York, 1998; pp 11−14. (b) Partington, J. R. *A History of Chemistry*; Macmillan & Co., Ltd.: London, 1970; Vol. 1, pp 55−62. (c) Pullman, B. *The Atom in the History of Human Thought*; Reisinger, A., Translator; Oxford University Press: New York, 1998; Chapter 4.

22. Plato. *Timaeus*; Jowett, B., Translator; http://classics.mit.edu/Plato/timaeus.html (accessed February 2, 2010).

23. (a) Partington, J. R. *A History of Chemistry*; Macmillan & Co., Ltd.: London, 1970; Vol. 1, Chapter 4. (b) Pullman, B. *The Atom in the History of Human Thought*; Reisinger, A., Translator; Oxford University Press: New York, 1998; pp 59−68.

24. Lucretius. *De Rerum Natura (On the Nature of Things)*; Leonard, W. E., Translator; http://classics.mit.edu/Carus/nature_things.html (accessed February 1, 2010).

25. (a) Partington, J. R. *A History of Chemistry*; Macmillan & Co., Ltd.: London, 1970; Vol. 1, Chapter 6. (b) Pullman, B. *The Atom in the History of Human Thought*; Reisinger, A., Translator; Oxford University Press: New York, 1998; pp 37−47.

26. Partington, J. R. *A History of Chemistry*; Macmillan & Co., Ltd.: London, 1970; Vol. 1, pp 194–195.

27. Gregory, J. C.*The Atom in the History of Human Thought*; Reisinger, A., Translator; Oxford University Press: New York, 1998; pp 20–21.

28. Gregory, J. C. *The Atom in the History of Human Thought*; Reisinger, A., Translator; Oxford University Press: New York, 1998; p 24.

29. (a) Stillman, J. M. *The Story of Alchemy and Early Chemistry*; Dover Publications, Inc.: New York, 1960; pp 136–143. (b) Taylor, F. S. *The Alchemists*; Barnes & Noble Books: New York, 1992; pp 68–69.

30. (a) Maimonides, M. *The Guide for the Perplexed*; Friedlander, M., Translator; George Routledge & Sons, Ltd.: London, 1947; Pt. 1, Chapters 71, 73. (b) Pullman, B. *The Atom in the History of Human Thought*; Reisinger, A., Translator; Oxford University Press: New York, 1998; Chapter 11.

31. Holmyard, E. J. Introduction. In *Ordinall of Alchimy*; Norton, T.; The Williams & Wilkens Co.: Baltimore, 1929; p iii, http://www.rexresearch.com/norton/norton.htm (accessed February 1, 2010).

32. Leicester, H. M. *The Historical Background of Chemistry*; John Wiley & Sons, Inc.: New York 1956; p 80.

33. Norton, T. *Ordinall of Alchimy*; The Williams & Wilkens Co.: Baltimore, 1929; p 79.

34. Pullman, B. *The Atom in the History of Human Thought*; Reisinger, A., Translator; Oxford University Press: New York, 1998; Chapter 9.

35. (a) Gregory, J. C. *The Atom in the History of Human Thought*; Reisinger, A., Translator; Oxford University Press: New York, 1998; p 23. (b) Leicester, H. M. *The Historical Background of Chemistry*; John Wiley & Sons, Inc.: New York 1956; p 110.

36. Partington, J. R. *A History of Chemistry*; Macmillan & Co., Ltd.: London, 1970; Vol. 2, p 386.

37. (a) Partington, J. R. *A History of Chemistry*; Macmillan & Co., Ltd.: London, 1970; Vol. 2, pp 381–383. (b) Pullman, B. *The Atom in the History of Human Thought*; Reisinger, A., Translator; Oxford University Press: New York, 1998; pp 132–136.

38. (a) Partington, J. R. *A History of Chemistry*; Macmillan & Co., Ltd.: London, 1970; Vol. 2, pp 271–276. (b) Leicester, H. M. *The Historical Background of Chemistry*; John Wiley & Sons, Inc.: New York 1956; p 111. (c) Newman, W. R. *Atoms and Alchemy, Chymistry and the Experimental Origins of the Scientific Revolution*; The University of Chicago Press: Chicago, 2006; Chapters 4, 5.

39. Partington, J. R. *A History of Chemistry*; Macmillan & Co., Ltd.: London, 1970; Vol. 2, pp 387–388.

40. (a) Partington, J. R. *A History of Chemistry*; Macmillan & Co., Ltd.: London, 1970; Vol. 2, pp 394–396. (b) Rees, G. *Ann. Sci.* **1980**, 549–571. (c) Cintas, P. *Bull. Hist. Chem.* **2003**, 65–75.

41. (a) Kubbinga, H. H. *J. Mol. Struct. (Theochem)* **1988**, *181*, 205–218. (b) Kubbinga, H. H. *J. Chem. Educ.* **1989**, *66*, 33.

42. (a) Pullman, B. *The Atom in the History of Human Thought*; Reisinger, A., Translator; Oxford University Press: New York, 1998; pp 122–132. (b) Shea, W. R. *Ambix* **1970**, *17*, 13–27.

43. (a) Gregory, J. C. *The Atom in the History of Human Thought*; Reisinger, A., Translator; Oxford University Press: New York, 1998; pp 31–32. (b) Partington, J. R. *A History of Chemistry*; Macmillan & Co., Ltd.: London, 1970; Vol. 2, pp 458–460. (c) Pullman, B. *The Atom in the History of Human Thought*; Reisinger, A., Translator; Oxford University Press: New York, 1998; pp 122–1132.

44. (a) Gregory, J. C. *The Atom in the History of Human Thought*; Reisinger, A., Translator; Oxford University Press: New York, 1998; pp 24–26. (b) Partington, J. R. *A History of Chemistry*; Macmillan & Co., Ltd.: London, 1970; Vol. 2, pp 430–441. (c) Pullman, B. *The Atom in the History of Human Thought*; Reisinger, A., Translator; Oxford University Press: New York, 1998; pp 157–161.

45. (a) Gregory, J. C. *The Atom in the History of Human Thought*; Reisinger, A., Translator; Oxford University Press: New York, 1998; p 24. (b) Leicester, H. M. *The Historical Background of Chemistry*; John Wiley & Sons, Inc.: New York 1956; pp 116–117. (c) Stillman, J. M. *The Story of Alchemy and Early Chemistry*; Dover Publications, Inc.: New York, 1960; pp 398–399.

46. (a) Gregory, J. C. *The Atom in the History of Human Thought*; Reisinger, A., Translator; Oxford University Press: New York, 1998; pp 32–37. (b) Newman, W. R. *Atoms and Alchemy, Chymistry and the Experimental Origins of the Scientific Revolution*; The University of Chicago Press: Chicago, 2006; Chapters 4, 5. (c) Partington, J. R. *A History of Chemistry*; Macmillan & Co., Ltd.: London, 1970; Vol. 2, Chapter 14. (d) Pullman, B. *The Atom in the History of Human Thought*; Reisinger, A., Translator; Oxford University Press: New York, 1998; pp 136–142. (e) Stillman, J. M. *The Story of Alchemy and Early Chemistry*; Dover Publications, Inc.: New York, 1960; pp 393–398.

47. Newton, I. *Opticks*; 1774; http://web.lemoyne.edu/~giunta/newton.html (accessed December 30, 2009).

48. (a) Partington, J. R. *A History of Chemistry*; Macmillan & Co., Ltd.: London, 1970; Vol. 2, pp 474–475. (b) Pullman, B. *The Atom in the History of Human Thought*; Reisinger, A., Translator; Oxford University Press: New York, 1998; pp 136–140.

49. (a) Gregory, J. C. *The Atom in the History of Human Thought*; Reisinger, A., Translator; Oxford University Press: New York, 1998; p 62. (b) Pullman, B. *The Atom in the History of Human Thought*; Reisinger, A., Translator; Oxford University Press: New York, 1998; pp 175–177.

50. (a) Gregory, J. C. *The Atom in the History of Human Thought*; Reisinger, A., Translator; Oxford University Press: New York, 1998; p 153. (b) Stillman, J. M. *The Story of Alchemy and Early Chemistry*; Dover Publications, Inc.: New York, 1960; pp 511–513.

51. (a) Gregory, J. C. *The Atom in the History of Human Thought*; Reisinger, A., Translator; Oxford University Press: New York, 1998; p 82. (b) Partington,

J. R. *A History of Chemistry*; Macmillan & Co., Ltd.: London, 1970; Vol. 3, pp 727–736.

52. (a) Gregory, J. C. *The Atom in the History of Human Thought*; Reisinger, A., Translator; Oxford University Press: New York, 1998; pp 67–74 . (b) Partington, J. R. *A History of Chemistry*; Macmillan & Co., Ltd.: London, 1970; Vol. 3, pp 736–749.

53. Higgins, W. *Experiments and Observations on the Atomic Theory and Electrical Phenomena*; 1814; http://www.google.com/books?id=ERkAAAAAQAAJ (accessed December 15, 2009).

54. (a) Gregory, J. C. *The Atom in the History of Human Thought*; Reisinger, A., Translator; Oxford University Press: New York, 1998; pp 74–76. (b) Partington, J. R. *A History of Chemistry*; Macmillan & Co., Ltd.: London, 1970; Vol. 3, pp 749–754.

55. Lavoisier, A.-L. *Elements of Chemistry*; Kerr, R., Translator; Dover Publications, Inc.: New York, 1965.

Chapter 4

150 Years of Organic Structures

David E. Lewis*

Department of Chemistry, University of Wisconsin-Eau Claire,
Eau Claire, WI 54702
*lewisd@uwec.edu

In 1858, a pair of papers were published that were to change the
way that organic chemists thought about the compounds they
dealt with. The first to appear was a paper by Friedrich August
Kekulé and the second, which appeared only slightly afterward,
was by a brilliant young Scotsman, Archibald Scott Couper.
Three years later, Aleksandr Mikhailovich Butlerov presented
his own form of the theory in a more usable form, and used
it not only to rationalize the chemistry of known compounds,
but to predict the existence of new compounds. Unlike modern
structural theory, none of the principals involved thought of
the structures as having a physical meaning, but, instead,
made a clear distinction between the chemical structure of the
compound, which could be deduced from its bonding affinities,
and its physical structure, which could not.

Introduction

The year 2008 was the sesquicentennial of a major milestone in the
development of organic chemistry. Separated by just a few weeks in 1858,
two papers appeared, setting out the basic principles of what is known today
as the structural theory of organic chemistry. The first paper, chronologically,
was by a young professor at the beginning of his independent career, Friedrich
August Kekulé (1829-1896). This paper (*1*), "Ueber die Constitution und die
Metamorphosen der chemischen Verbindungen, und über die chemische Natur
des Kohlenstoffs," was received by the editor of the *Annalen der Chemie und
Pharmacie* on March 16, 1858, and published in May that year. The second to
appear was a paper by a young Scotsman, Archibald Scott Couper (1831-1892),
who was a student working in the Paris laboratory of Charles Adolphe Wurtz

(1817-1884). This paper (*2*), "Sur une nouvelle théorie chimique," was eventually presented before the Académie des Sciences by Dumas, and published in the *Comptes rendus* on June 14, 1858, six weeks after Kekulé's paper appeared. Wurtz was not a member of the Académie at that time, and therefore could not present the paper himself without the sponsorship of a member (usually Balard). Whatever its cause, Wurtz' delay in finding a sponsor brought Couper to a fury, with far-reaching consequences.

In both papers two key ideas were set forth — 1) the idea of the tetravalent carbon atom, and 2) the concept of catenation (that carbon atoms could form chains). Both papers also stated explicitly, for the first time, what became known as the chemical structure of a compound, and the idea that the properties of the compounds depended on the properties and arrangement of their component atoms, rather than the more complex (and less well defined) radicals. The appearance of these papers also resulted in a polemical interchange over priority that left Couper a broken man, and boosted Kekulé (with the help of Wurtz, amongst others) to the top echelons of organic chemistry in Europe (*3*).

Organic Chemistry Prior to 1858: Radicals, Types, and Substitution

Before we proceed with a discussion of these two papers, it is worthwhile examining the state of organic chemistry in 1858 – a mere 30 years after Wöhler's serendipitous synthesis of urea while attempting to prepare ammonium cyanate (*4*). These three decades had been eventful, seeing, amongst other things, the first total synthesis of an organic compound from its elements — acetic acid, by Hermann Kolbe (1818-1884) in 1845 (*5*) — and the preparation of the first organometallic compound — diethylzinc, by Edward Frankland (1825-1899) in 1852 (*6*). (Although Bunsen's cacodyl (*7*) might be viewed as an organometallic compound, which would make it the first organometallic compound, arsenic is a metalloid rather than a true metal like zinc, which makes the claim for diethylzinc as the first "truly" organometallic compound tenable, at least.) At the same time, progress in organic chemistry was still hampered by the lack of uniformity in atomic weights during these early years, a situation that was eventually settled by Stanislao Cannizzarro (1826-1910) in the Karlsruhe Congress of 1860 (*8*). Thus, during its developmental stages, much of organic chemistry was described in terms of equivalent weights, with oxygen having an equivalent weight of 8, and carbon an equivalent weight of 6. This necessitated the use of "double atoms" of these elements when writing molecular formulas, and barred symbols were frequently used to represent these atoms. The distinction between an atom and a molecule had begun with Dalton ((*9*), although he did not use those explicit terms), and the distinction between atoms and molecules in gaseous elements had been made by Amadeo Avogadro (1776-1856) (*10*) and André Marie Ampère (1775-1836) (*11*) before their reiterations in 1833 by Ampère's student, Marc Antoine Augustin Gaudin (1804-1880) (*12*), and 1846 by Auguste Laurent (1807-1853) (*13*); Frankland had introduced the concept of valence in 1852 (*14*).

The first major theory of organic chemistry to follow Wöhler's urea synthesis was his establishment, with his friend Justus von Liebig (1803-1873), of the concept of the multi-atom organic radical – benzoyl, described in the landmark paper of 1832 (*15*). The definition of a radical was further clarified by Liebig in his 1837 paper, where he defined a radical in terms of three major characteristics (*16*):

"We call cyanogen a radical (1) because it is a non-varying constituent in a series of compounds, (2) because in these latter it can be replaced by other simple substances, and (3) because in its compounds with a simple substance, the latter can be turned out and replaced by equivalents of other simple substances."

In the decade after Liebig and Wöhler established the multi-atom radical as a presence in organic chemistry,the next logical step was being taken by Laurent (*17*), Jean-Baptiste André Dumas (1800-1884) (*18*), and Charles Frédéric Gerhardt (1816-1856) (*19*), who were developing substitution theory, and its logical offshoot, type theory (*20*). Dumas, especially, had been struck by the fact that trichloroacetic acid was, in almost all respects, chemically analogous to acetic acid, and he suggested that one might be able to substitute hydrogen in an organic substance by halogen without appreciably affecting its chemistry. He confirmed this by studies wherein he chlorinated oil of turpentine (*21*) and alcohol (*22*). This was at odds with the dualistic theory of Berzelius (*23*) and with the experience of inorganic chemistry, where the substitution of hydrogen by halogen usually gave rise to dramatic changes in its chemical properties (one need only compare sodium chloride with sodium hydride). Dumas also expanded this suggestion to other elements, noting the similarities in the properties of potassium permanganate and potassium perchlorate. Substitution theory did not sit well with the German schools of organic chemistry, however, and this led to the famous (infamous) parody written in French by Wöhler, under the pseudonym S. C. H. Windler (i.e. *Schwindler*, a swindler or mountebank), in which every atom of manganous acetate was replaced by chlorine — a *reductio ad absurdum* of Dumas' ideas (*24*). Within a decade, however, the work of Alexander William Williamson (1824-1904), whose work defined the "water type," (*25*) and August Wilhelm von Hofmann (1818-1892), whose work on what he called the volatile bases defined the "ammonia type," (*26*) had produced results that validated type theory, and allowed Gerhardt to publish what became known as the "Newer Type Theory," (*27*) which today we would probably describe as classification according to functional group.

1858 – The Pivotal Year

By 1858, the stage was set for the emergence of structure theory. This involved, as pointed out earlier, a pair of critical insights into the bonding in carbon compounds — what we now call the tetravalence of carbon, and its propensity for catenation. It should be pointed out at this stage, also, that neither

Kekulé nor Couper intended the structures they wrote to be interpreted as having any significance as a *physical* representation of the molecule. Like Butlerov after them, they were careful to distinguish between *physical* and *chemical* structure. The caveat implied here was that the structures drawn indicated the *chemical* locations of the atoms, and may or may not correspond to the physical positions of those same atoms in the molecule; atoms close together "chemically" were not necessarily close together physically. In other words, the structures were actually maps of the linkages between the "affinities" of the individual atoms in the molecule.

Like their contemporaries, none of the major protagonists in the early debates about chemical structure believed that it was possible (at that time, at least) to determine or deduce the physical locations of atoms in a molecule or radical. In some ways, therefore, this clear distinction between the chemical and physical structure of the molecule may have been a way to temper criticism of the new ideas by their more conservative contemporaries, and thus render the concept more palatable to a wider audience of chemists.

Friedrich August Kekulé

The first paper to appear during the pivotal year of 1858 was Kekulé's paper in the *Annalen der Chemie und Pharmacie*. Kekulé, shown in Figure 1, was born in Darmstadt. From 1848-1851 he studied architecture at Giessen. Here he came under the influence of Liebig, who persuaded him to change his focus to chemistry. He followed his education at Giessen with studies in Paris (1851-1852) and London (1854-1855). In Paris he studied under August André Thomas Cahours (1813-1891), Charles Adolphe Wurtz (1817-1884), and–significantly–Gerhardt; in London, he worked in the St. Bartholomew's Hospital laboratory of John Stenhouse (1809-1880), and met and befriended Williamson and William Odling (1829-1921). It was during his time in London that he began to develop his theory of molecular architecture (structure theory). On his return to the continent, he worked first as a Privatdozent in Heidelberg, and then, in 1858, he moved to Ghent as professor. In 1867, he moved to a chair at Bonn, and remained there the rest of his life.

Kekulé had actually first set out the concept of the tetravalence of carbon explicitly in a footnote in a paper published in 1857: "Carbon is, as may easily be shown, and as I shall explain in greater detail on a later occasion, tetrabasic or tetratomic; that is, 1 atom of carbon = C = 12, is equivalent to 4 at. H. The simplest compound of carbon with an element of the first group, with H or Cl for example, is accordingly: CH_4 and CCl_4" (*28*). In his 1858 paper, Kekulé expanded on this idea, and set out explicitly what he intended his theory to accomplish: "I regard it as necessary and, in the present state of chemical knowledge, as in many cases as possible, to explain the properties of chemical compounds by going back to the elements themselves which compose those compounds. I no longer regard it as the chief problem of the time, to prove the presence of atomic groups which, on the strength of certain properties, may be regarded as radicals... I hold that we must extend our investigation to the constitution of the radicals themselves... As my starting point, I take the views I developed earlier concerning the nature of the elements and the basicity of atoms" (29).

In his 1858 paper, Kekulé also set forth a notation for representing organic reactions, in which the reactants and products were shown in the form of type formulas. In Kekulé's notation, the reactants were arranged vertically on the left, and the products were arranged vertically on the right. Figure 2 illustrates this notation for the hydrolysis of succinimide to succinamic acid, along with an interpretation of the original Kekulé formulas using lines to indicate bonds and with the modern representation of the same reaction using full structural formulas.

Kekulé's paper specifies four types of reactions, although much of the body of the paper focuses on metathesis (double decomposition) reactions, implying that they are most important. In the example above, the reaction is written as just such a reaction. Kekulé's paper specifies, for the first time, that the atoms in radicals are bound together in an explicit arrangement (in terms of their chemical affinities), although he did not take the further step to specify that each chemical substance has a single, unique structure. That was left to the Russian, Butlerov.

The clearest statement of the importance of structure comes towards the end of Kekulé's paper, in a section entitled, "Constitution der Radikale. Natur des Kohlenstoffs. [The constitution of radicals and the nature of carbon]." In this section, Kekulé clearly expounds the ability of carbon to bind to itself, making the point that carbon is tetravalent (in his terminology, it has four chemical [i.e. affinity] units) so a C_2 unit is hexavalent, and a C_5 unit has 12 chemical units, and so forth; at the end of this same paragraph, he deduces the $2n+2$ rule for chemical units associated with n carbon atoms. He then makes the following important point: "When, comparisons are made between compounds which have an equal number of carbon atoms in the molecule and which can be converted into each other by simple transformations (e.g. alcohol, ethyl chloride, acetaldehyde, acetic acid, glycolic acid, oxalic acid, etc.) the view is reached that the carbon atoms are arranged in the same way and only the atoms held to the carbon skeleton are changed" (29). What is truly remarkable about this paper, given the ferocity with which Kekulé later defended his claim to priority in developing structural theory, is the way in which he dismisses it in the final paragraph: "Lastly, I feel bound to emphasize that I myself attach but a subordinate value to considerations of this kind... it appears appropriate that these views should be published, because they seem to me to furnish a simple and reasonably general expression precisely for the latest discoveries, and because, therefore, their application may perhaps conduce to the discovery of new facts" (29).

Archibald Scott Couper

Archibald Scott Couper was born in Kirkintilloch, Scotland, and educated at home due to his poor health as a child. He followed his early education with study at the University of Glasgow, where he studied the humanities and languages, and at the University of Edinburgh, where he studied logic, metaphysics, and moral philosophy. At some time between 1854 and 1857, he decided to study chemistry, and he moved to Paris, where he entered the laboratory of Wurtz. Figure 3 shows Couper during his time in Paris. In the space of one year, he published

Figure 1. Friedrich August Kekulé (1829-1896)

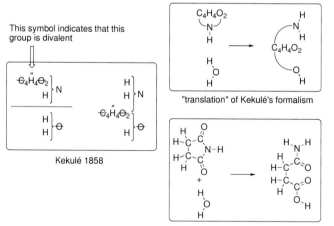

Figure 2. Notation for organic reactions used by Kekulé in his 1858 paper for the hydrolysis of succinimide to succinamic acid.

four papers, the first describing his work on the bromination of benzene (*30*), the second on reactions of salicylic acid with phosphorus pentachloride (*31*), and his two papers on structural theory. Couper's reaction to the delay in finding a presenter for his paper on structural theory was not temperate, and it resulted in a serious falling out with Wurtz, and his summary expulsion from Wurtz' laboratory. (In a letter to Richard Anschütz, Albert Ladenburg describes Wurtz as having "bungled this a little" (*32*).) In 1859, he returned to Scotland, where he joined the laboratory of Lyon Playfair, but a nervous breakdown shortly thereafter ended his scientific career, and his contributions to structural theory were all but forgotten. Indeed, Couper would have been forgotten entirely had not his contributions been rediscovered by (ironically!) Kekulé's successor at Bonn, Richard Anschütz (*33*).

It is ironic that in his paper presented to the Académie des Sciences , Couper makes a statement almost identical to one that occurs in Kekulé's 1858 paper: "I go back to the elements themselves, of which I study the mutual affinities. This study is, in my opinion, sufficient for the explanation of all chemical combinates, without it being necessary to revert to unknown principles and to arbitrary generalizations." Couper's ultimate goal is most clearly spelled out in the opening paragraph of his paper in the *Philosophical Magazine and Journal of Science* (*34*), where he states: "The end of chemistry *is its theory*. The guide in chemical research *is a theory*. It is therefore of the greatest importance to ascertain whether the theories at present adopted by chemists are adequate to the explanation of chemical phenomena, or are, at least, based upon the true principles which ought to regulate scientific research."

Unlike Kekulé's paper, which was careful to maintain a connection with the earlier work on the theory of Types, and which described an incremental advance in the theory of chemical structure, Couper's papers completely broke ranks with Gerhardt's theory of types, and he deliberately set out in his papers to illustrate the shortcomings of type theory. This he did by using a linguistic analogy to make the point that type theory does not address the *fundamental* problem of organic chemistry, and declared instead that it was important to return to first principles when discussing organic compounds. Applying this method of analysis to carbon, he arrived at two critical fundamental principles:

"– 1° It combines with equal numbers of equivalents of hydrogen, chlorine, oxygen, sulfur, etc., which can be mutually replaced to satisfy its power of combination.
– 2° It enters into combination with itself.
"These two properties in my opinion explain all that is characteristic of organic chemistry. I believe that the second is specified here for the first time. In my opinion, it accounts for the important, and still unexplained, fact of the accumulation of molecules [atoms] of carbon in organic compounds. In compounds 2, 3, 4, 5, 6, etc., molecules [atoms] of carbon are linked together..." .

Couper's paper also differed from Kekulé's in another important way: it was illustrated with graphic formulas of molecules, with bonds between atoms explicitly shown by a series of dotted lines. In the more extensive follow-up

Figure 3. Archibald Scott Couper (1831-1892)

paper with the same title published later the same year (*35*), Couper replaced the dotted lines with dashes to give structural formulas that clearly presage the modern versions, as shown in Figure 4 for methanol, ethanol, and cyanuric acid.

The last of these three examples is important, since it shows an explicit representation of a cyclic compound almost a decade before Kekulé proposed his cyclic formula for benzene. This last formula is also important in showing that Couper did not explicitly represent double bonds between atoms in his structures.

It may also have been unfortunate for Couper that he was an excellent experimentalist, because his greater skill in the laboratory unwittingly gave ammunition to his opponents. In his early work that culminated in structural theory, he reported the formation of a cyclic compound from the reaction between phosphorus pentachloride and salicylic acid (*31*). Both Kekulé (*36*) and Kolbe (*37*) attempted to reproduce his reaction without success, and used their failure to duplicate his results to cast doubt on his other claims. It was not until a better experimentalist, Richard Anschütz, repeated Couper's work *exactly as he had set out in his paper* that his skill as an experimentalist was confirmed, and his reputation was restored (*38*). Kekulé and Kolbe had both continued heating the reaction beyond the time specified by Couper, and therefore obtained the acid chloride. Couper's skill was again verified by Pinkus and his coworkers in the 1960s (*39*). When looked at in the light of his career – he was a young man of 27 years with only two to three years of formal training in chemistry – Couper's papers are remarkable, and one can only wonder what he might have produced had that career not been cut short.

Refinement of the Theory of Chemical Structure: 1859-1870

Following the publication of structural theory for all to examine, the next major players in its development were the Russian, Aleksandr Mikhailovich Butlerov (1828-1886), the Scot, Alexander Crum Brown (1838-1922), and the Austrian, Johann Josef Loschmidt (1821-1895).

Aleksandr Mikhailovich Butlerov (Butlerow, Boutlerov)

Immediately upon its publication, Couper's work came under fire. The first to attack his theory was Kekulé, who summarily claimed priority for the ideas contained therein in a paper entitled, "Remarques par M. A. Kekulé à l'occasion d'une Note de M. Couper sur une nouvelle théorie chimique" (*40*) – even though he may have misunderstood (*41*) part of Couper's formulas and symbolism. Wurtz, too, found fault with Couper's paper, calling his structural formulas "too arbitrary and too far removed from experiment" (*42*); he found fault with his use of language, which he considered intemperate, his use of the atomic weight of 8 for oxygen, and his summary dismissal of the established theories of organic chemistry. As a consequence of this (and it is tempting to speculate that it was also a result of his rancorous parting with Couper), Wurtz did not advance Couper's claims as a co-creator of structural theory (*43*) in favor of Kekulé's claims, although he did eventually acknowledge Couper as an independent co-inventor of structural theory in a footnote in his history of chemical theories (*44*). The other major criticism of Couper's paper came from Butlerov (*45*), who considered Couper's ideas premature and not supported by the available evidence. As we shall see below, Butlerov's opinion changed with time.

Butlerov, whose portrait is shown in Figure 5, is a figure whose place in the development of structural theory has been clouded by politics (*46*). Born to a family in the lesser nobility, Butlerov was educated at Kazan' University, where he studied under Nikolai Nikolaevich Zinin (whose discoveries include the reduction of nitrobenzene to aniline) and Karl Karlovich Klaus (the discoverer of ruthenium). Although his thesis for *kandidat* (approximately a modern Ph.D.) was on the butterflies of the Volga-Ural region (*47*), Zinin's departure for St. Petersburg left the University needing a junior colleague for Klaus to teach chemistry, so Butlerov was appointed to the post. Following Klaus' departure for Dorpat (now Tartu, in Estonia), the sole responsibility for teaching chemistry at Kazan' University devolved upon Butlerov. During a *komandirovka* (study leave) in western Europe, he became acquainted with both Kekulé and Couper,

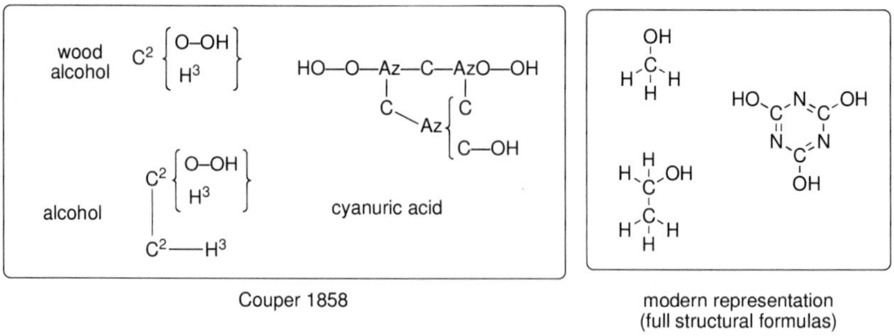

Figure 4. Structural formulas from Couper's second paper in 1858, and the modern versions of the same formulas. Note that Couper's structure for cyanuric acid is not correct, having the oxygen atoms bonded to nitrogen.

and he quickly became a disciple of the new ideas being promulgated among the chemists of the younger generation. On his return to Russia, he continued to develop his version of the theory, and he wrote the first textbook of organic chemistry based exclusively on structural theory (*48*). Butlerov was an inspiring teacher who was trusted and admired by students and faculty alike: he served two terms as Rector of Kazan' University (the second at the insistence of the students) before moving to St. Petersburg, where he spent the remainder of his career.

During the Soviet era, Butlerov was vigorously championed by Russian historians of science and some in the west as the true creator of structural theory (*49*). Unfortunately, this effort to give Butlerov priority in *creating* the theory meant that his real, *and critically important*, contributions in *developing* the theory were largely overlooked. An interesting analogy can be found in the Wittig reaction: Although the reaction between a phosphorane and a carbonyl compound had been discovered by Hermann Staudinger (*50*) in 1919, it was the work of Georg Wittig over three decades later (*51*) that led to it becoming a truly useful synthetic method. Likewise, although Couper and Kekulé had created structural theory, it was Butlerov who best understood the true potential of their theory, and who made their creation useful.

Butlerov was the first to use the explicit term, "chemical structure," and he presented structural theory in a much more useful form than either Kekulé's or Couper's papers (*52*), showing organic chemists how to get the maximum benefit from its application. It was Butlerov's explicit statement that each compound is represented by a single chemical structure, and that each chemical structure represents a single compound, that was the defining statement of his version of the theory. This went further than either Couper or Kekulé had done, and encouraged the use of structural theory in predicting, for example, the number and structures of isomers of organic compounds. Over the next several years, Butlerov did just that (*53*); in 1876, he used structural theory to provide the first explanation of the phenomenon that later became known as tautomerism (*54*). More importantly, he verified his structural predictions with validating syntheses of compounds such as *tert*-butyl alcohol (*55*) and isobutylene (*56*). Moving structural theory from an organizing tool to a predictive tool was critical to its wider acceptance.

Figure 5. Aleksandr Mikhailovich Butlerov (1828-1886)

Butlerov's Version of Structural Theory Expounded: The Speyer Paper

Butlerov presented his first paper on structural theory, "Einiges über die chemische Struktur des Körpers," to the Chemistry Section at the Congress of German Naturalists and Physicians at Speyer in 1861 (*52*). In it, he stated, "The well known rule that says that the nature of compound molecules depends on the nature, the quantity, and the arrangement of its elementary constituents can for the present be changed as follows: the chemical nature of a compound molecule depends on the nature and quantity of its elementary constituents and on its chemical structure." Although Butlerov had criticized Couper's structures in 1859, at the end of his Speyer lecture in 1861 he made the following telling comment: "I am even obliged to remark that the theory and formulas of Couper—whose too absolute and exclusive conclusions I disputed at that time—contained similar thinking. It was, however, neither clearly enough perceived nor expressed."

Emil Erlenmeyer (1825-1909), the editor of the *Zeitschrift für Chemie*, understood Butlerov's models better than most of his contemporaries, and published the paper in his journal. To him is due much of the credit for the rise of popularity of Butlerov's ideas.

The formulas in Butlerov's paper, like those of Couper, showed bonds between groups of atoms, but did not show every bond in a molecule explicitly; his formulas retained some vestiges of Type formulas, and tend to resemble modern condensed structural formulas. Butlerov was, however, fully committed to the new structural theory, although he did recognize that for some compounds it was not necessary to specify every bond explicitly.

Alexander Crum Brown

The move towards the explicit designation of every bond in a molecule, and modern graphical formulas, was the result of the work of Alexander Crum Brown (shown in Figure 6), professor of chemistry at the University of Edinburgh. Crum Brown was born and raised in Edinburgh, taking his M.D. from the University of Edinburgh in 1861, and his D.Sc. from the University of London (the first D.Sc. degree conferred by the University) in 1862. He spent the year following his graduation from London in Germany, working with Bunsen and Kolbe, and he then returned to Edinburgh to a junior faculty position in 1863. He succeeded Lyon (later Lord) Playfair to the Chair of Organic Chemistry in 1869, being chosen in preference to other candidates such as William Henry Perkin, Sr., of mauve fame. Interestingly, one of his letters of support was written by Butlerov. Crum Brown's intellect was powerful and wide-ranging, and his work was of high caliber, but due to his propensity for publishing his work predominantly in the *Transactions of the Royal Society of Edinburgh*, and the corresponding *Proceedings*, much of his work was not well known by the chemistry community at large: it was neither widely disseminated nor widely read. His interests ranged from chemistry and mathematics (most of his contributions to organic chemistry were theoretical and mathematical in content) to physiology, philosophy and church history. While he is reported (*57*) to have had a knowledge of several languages, including Russian and Chinese, and to have served as Examiner in Japanese, his command of Russian may not have been to the level of fluency, since his correspondence with Butlerov was in French, and not Russian. Nevertheless, he was reputed by his contemporaries to be capable of "filling any Chair in the University."

In his M.D. thesis, presented in 1861, Crum Brown drew structures of molecules that are remarkably like the modern formulas (*58*), with the exception that he drew the symbol of each element inside a circle (reminiscent of Dalton's symbolism), and he drew double bonds with bent lines, rather than straight lines, as shown in Figure 7, for succinic acid. He followed this work up with a paper entitled, "On the classification of chemical substances by means of generic radicals" (*59*).

The obvious resemblance between Crum Brown's formulas and those of Couper, whose brief time at Edinburgh after his return from France overlapped with Crum Brown's student days, raises the equally obvious question of whether or not Crum Brown was acquainted with Couper's formalism for writing chemical structures. His biographer in the *Journal of the Chemical Society* maintains (*60*) that although both men were in Playfair's laboratory in 1858 (Couper had returned to Edinburgh in late autumn, 1858, and was offered the position in Playfair's laboratory that December) Crum Brown was not, in fact, acquainted with Couper's work.

Two letters from Frankland are quoted in Crum Brown's obituary (*60*), and provide some insights into the reception of his graphical representations. On May 28, 1866, Frankland wrote, "I am much interested in graphic formulae and consider that yours have several important advantages over Kekulé's. In my lectures here last autumn I used them throughout the entire course and with

Figure 6. Alexander Crum Brown (1838-1922)

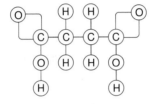

Figure 7. Crum Brown's structural formula for succinic acid.

very great advantage, and I now have in the press a little book of Lecture Notes for Chemical Students in which they are copiously used." In the second letter, written on June 4, 1866, Frankland notes, "I am just now endeavouring to get Kolbe to express certain of his fundamental formulae graphically. We should then understand each other better. There is a good deal of opposition to your formulae here, but I am convinced that they are destined to introduce much more precision into our notions of chemical compounds. The water-type, after doing good service, is quite worn out." Prophetic words, indeed!

Johann Josef Loschmidt

The third innovator of the early 1860s was the Austrian physicist Johann Josef Loschmidt, who, despite his work in chemistry, is not widely appreciated as a chemist. A portrait of him is shown in Figure 8. Loschmidt was born in Pocena and began his university education at the German University of Prague. He completed his education at Vienna, but was unable to procure a teaching position there. He moved into industry, but became involved in a series of financially unsuccessful (although technically sound) ventures; he eventually became a concierge at the University of Vienna before qualifying as a high school teacher. He later became Privatdozent at Vienna, and ended his career there as Professor of Physics – in the same university where he had been concierge!

Loschmidt is remembered as a physicist for his determination of the Loschmidt number (*61*), the number of molecules in 1 mL of gas, but in 1861 he published a small booklet (*62*) in which he drew structural formulas for a large number of organic compounds. Two formulas of the Loschmidt type are reproduced in Figure 9. It is interesting to note that, in many ways, Loschmidt's formulae reflect Dalton's elemental symbols, and may, in fact, be based on Dalton's symbolism.

It has been argued (*63*) that this pamphlet contains the first correct structure for benzene, but this is an over-interpretation of Loschmidt's structures for benzenoid aromatic compounds. In his formulas, he represented most elements by circles whose radii were proportional to their atomic weights (where the atomic weights were close, he differentiated the atoms by the number of concentric circles – carbon was a single circle, oxygen was a double circle, and nitrogen was a triple circle, as shown in Figure 9). In like manner, he represented the hexavalent C_6 unit of the benzene nucleus by a large circle with an atomic weight of 78, and he then distributed the other groups around this large circle. Where the available evidence suggested that the groups were adjacent to each other, he placed them in adjacent positions on the C_6 ring. What is not arguable about Loschmidt's formulas is his unequivocal and explicit representation of double and triple bonds between elements, as shown in Figure 9 for two aromatic compounds.

Years before Kekulé's benzene formula was published, Loschmidt represented what translates into the Kekulé structure of the aromatic ring of 1,3,5-triazine, and this may be what has led to the assertion that he obtained the structural formula for benzenoid aromatic compounds before Kekulé. The fact that he did not explicitly represent the six atoms of the benzenoid aromatic ring with alternating double and single bonds as he did for the triazine ring suggests that this is not, in fact the case. Regardless of this, it is unfortunate for Loschmidt that his system was cumbersome to use, so it found little favor with practicing chemists.

Chemical Structure Becomes *Physical* Structure (1874)

The gradual adoption of structural theory occurred with remarkably little resistance, although conservative chemists took more time to come to the more modern theoretical view. Hermann Kolbe, for example, held to his theory of "rational constitution," although by the end of the 1860s, the differences between Kolbe's view and structural theory had largely disappeared (*64*).

This acceptance brought with it the inevitable questions about whether the distinction between the chemical structure of a molecule and its physical structure was real. By the end of the 1860s, most of the vehement denials that the graphical formula of an organic compound held any physical significance—which had been so important a part of the early papers on structural theory—had become muted.

Figure 8. Johann Josef Loschmidt (1821-1895)

LOSCHMIDT 1861   MODERN FORMULAS

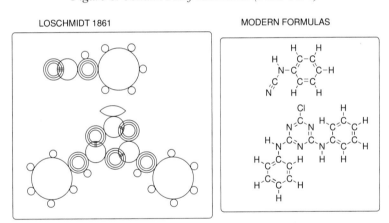

Figure 9. Loschmidt's structures for phenylcyanamide and the product formed from two equivalents of aniline and cyanuric chloride.

The start of the movement towards physical meaning for graphic formulas of organic compounds may be traced to Louis Pasteur (shown in Figure 10), whose work with the tartaric acids led to the proposal, presented in two lectures presented before the Société Chimique de Paris on January 20 and February 3, 1860 (*65*), that molecules of the same substance with opposite rotations might be related to each other as object to mirror image. The importance of configuration at the molecular level was reinforced by the work of Johannes Wislicenus (shown in Figure 10) (*66*), who suggested that the differing properties of stereoisomers could be traced to differing three-dimensional arrangements of their atoms.

Figure 10. Louis Pasteur (1822-1895) Johannes Wislicenus (1835-1902)

*Figure 11. Jacobus Henricus van't Hoff (1852-1911) Joseph Achille Le Bel
(1874-1930)*

The introduction of the tetrahedron into discussions of carbon compounds occurred in 1862, but when Butlerov suggested that year, that an asymmetric tetrahedral model with the individual affinities on the faces could be used to explain why different valences of carbon had different strengths, and rationalize the (spurious) difference between hydride of ethyl (C_2H_5—H) and ethane (CH_3—CH_3) (*67*), it was as a heuristic aid, and not to imply the physical structure of the molecule. In 1867, Kekulé used tetrahedral models of the carbon atom to rationalize the bonding in acetylene and hydrogen cyanide (*68*); a copy of the Butlerov paper carrying Kekulé's annotations has been discovered (*69*). Still, there is no evidence that models were intended to serve as anything but useful pedagogical devices; they were not intended to represent the actual *physical shape* of carbon atoms in organic molecules. That step came six years later.

By 1874, the Dutch chemist Jacobus Henricus van't Hoff (1852-1911) and the French chemist Joseph Achille Le Bel (1847-1930), working independently, had simultaneously arrived at basically the same conclusion – that the formulas written for organic compounds should be interpreted as having a *physical* significance as well as a chemical significance (i.e atoms that had been interpreted as being chemically close to each other were also physically close to each other). Figure 11 shows van't Hoff and Le Bel. This advance in structural theory appeared in the form of a small pamphlet published in Dutch by van't Hoff (*70*), and a paper by Le Bel, "Sur les relations qui existent entre les formules atomiques des corps organiques, et le pouvoir rotatoire de leurs dissolutions," published in the *Bulletin de la Société Chimique de France* (*71*). It is possible that van't Hoff, who was working with Kekulé at the time, may have been exposed to the idea of a non-planar carbon at the time of Kekulé's 1867 paper. This does not, however, imply that van't Hoff's tetrahedral carbon theory is not all of his own development: the seed may have come from his time with Kekulé, but the theory that emerged is van't Hoff's.

This time, there were no polemics and no fight over priority (neither of the principals had Kekulé's driving ambition to be the first and only inventor of the theory) – in fact, in his subsequent papers on stereochemistry, van't Hoff was careful to point out Le Bel's contributions, and may have saved them from obscurity. In both 1874 papers, the authors asserted that the physical properties of organic compounds, especially their optical activity, could be accounted for by specifying that the molecules contained an asymmetric atom corresponding to a tetrahedron surrounded by four different groups. While van't Hoff's paper concentrated on the tetrahedron as the basis for optical activity, Le Bel's paper was more wide-ranging, and allowed other chiral shapes to be considered.

There is a question of just how rapidly the concepts in van't Hoff's and Le Bel's papers would have been adopted by the organic chemistry community at large had not Johannes Wislicenus championed them. Wislicenus wrote the preface to the German translation of van't Hoff's booklet, which he had asked his student, F. Hermann, to translate into German. There is no doubt that having someone of Wislicenus' status as a champion helped the tetrahedral carbon gain acceptance, but one must also give Kolbe some credit. The dissemination of these ideas was almost certainly also helped by his (unfortunately for his legacy) intemperate, even legendary, tirade (*72*) against the tetrahedral carbon in van't Hoff's paper, since it gave them a wider readership than they might otherwise have enjoyed.

Conclusion

By the end of the 1860s, graphical formulas for organic compounds were so widely accepted that their use had become routine, although their translation to representations of the physical positions of the atoms took another half decade. The upshot of this rapid acceptance was that the progenitors of the theory, who had fought so furiously over priority claims, were not really given the full credit for their discoveries—when a theory becomes so "obvious" to all practitioners of a science, its novelty is forgotten, and few remember the hard work of, or feel it necessary to recognize, the pioneers. Even so, the controversy over priority for the development of modern graphical formulas and the structural theory of organic chemistry did not end with the nineteenth century. In 1923, Sir James Walker stated, in his Presidential Address to the Chemical Society of London, that the use of graphical formulas in organic chemistry could be traced from Couper through Crum Brown to Wurtz, and not through Kekulé and Loschmidt (73), and we have already discussed the efforts of Soviet historians of science to assign credit for the theory to Butlerov. It is an interesting footnote to history that in 1868, the year after Kekulé had left Ghent for Bonn, his former colleague, Carl Glaser, immediately adopted Erlenmeyer's structural formulas (74), which Erlenmeyer had, in turn, adapted from those of Crum Brown (75) — just ten years after Couper had formulated his structural theory, his structural formulas displaced Kekulé's in Kekulé's own former department, surely one of the ultimate ironies of the whole story.

It is remarkable that, in the space of less than two decades, the structural theory of organic chemistry should have moved from the first hesitant steps, where the chemical structure was considered to be separate and distinct from the physical structure of the molecule, and represented only the "affinities" of the atoms within the molecule, to the point where those same formulas were now viewed as representations of the actual physical locations of the atoms in the molecule (76). What was left undone at the end of the nineteenth century, by which time three-dimensional graphical formulas for organic compounds were in routine use, was, of course, a description of exactly what the "chemical affinities" of the atoms composing the molecules were. The answer to this problem would have to await the new century, and the development of modern theories of the atom and bonding.

References

1. Kekulé, A. *Ann. Chem. Pharm.* **1858**, *106*, 129–159.
2. (a) Couper, A. S. *Comptes Rend.* **1858**, *46*, 1157–1160. This paper was translated by Leonard Dobbin as part of a 1953 publication by the Alembic Club, *On a New Chemical Theory and Researches on Salicylic Acid. Papers by Archibald Scott Couper.* (b) Couper, A. S. *Ann. Chim.* [3] **1858**, *53*, 469–489. These papers appeared in English and German. (c) Couper, A. S. *Philos. Mag.* **1858**, *16*, 104–116. (d) Couper, A. S. *Ann. Chem. Pharm.* **1859**, *110*, 46–51.
3. For a lucid discussion of the events of May–June, 1858, see Rocke, A. *Nationalizing Science*; MIT Press: Cambridge, 2001; pp 207–211.
4. (a) Wöhler, F. *Pogg. Ann.* **1828**, *12*, 253–256. (b) Wöhler, F. *Ann. Phys. Chem.* **1828**, *37*, 330–333.
5. Kolbe, H. *Ann. Chem. Pharm.* **1845**, *54*, 145–188.
6. Frankland, E. *Quart. J. Chem. Soc.* **1850**, *2*, 297–299.
7. (a) Bunsen, R. W. *Ann. Pharm.* **1839**, *31*, 175. (b) Bunsen, R. W. *Ann. Chem. Pharm.* **1841**, *37*, 1. (c) Bunsen, R. W. *Ann. Chem. Pharm.* **1842**, *42*, 14. (d) Bunsen, R. W. *Ann. Chem. Pharm.* **1843**, *46*, 1.

8. (a) Cannizzaro, S. *Il Nuovo Cimento* **1858**, *7*, 321–366. Available in English translation as Cannizzaro, S. Sketch of a Course of Chemical Philosophy. In *Alembic Club Reprints*, No. 18; E. & S. Livingstone: Edinburgh, 1910. (b) Ihde, A. J. *J. Chem. Educ.* **1961**, *38*, 83–86.

9. Dalton, J. *Mem. Manch. Lit. Philos. Soc.* **1805**, *1*, 171.

10. Avogadro, A. *J. de Phys.* **1811**, *73*, 58.

11. Ampère, A. M. *Ann. Chim. Phys.* **1814**, *90*, 45.

12. Gaudin, M. Nouvelle manière d'envisager les corps gazeux, avec applications à la détermination du poids relatif des atomes. *Ann. Chim.* **1833**, *52*, 113, submitted to the Académie des Sciences October 30, 1831.

13. (a) Laurent, A. *Ann. Chim.* **1846**, *18*, 266–298. (b) Laurent, A. *J. Prakt. Chem.* **1847**, *40*, 65–89.

14. Frankland, E. *Philos. Trans. R. Soc. London* **1852**, *142*, 417–444.

15. Wöhler, F.; Liebig, J. *Ann. Chem. Pharm.* **1832**, *3*, 249–287.

16. Cited in Ladenburg, A. *Lectures on the History of the Development of Chemistry Since the Time of Lavoisier*; Dobbin, L., Translator; Alembic Club: Edinburgh, U.K., 1900.

17. (a) Laurent, A. *Ann. Chim. Phys.* **1832**, *50*, 182. (b) Laurent, A. *Ann. Chim. Phys.* **1836**, *61*, 125. (c) Laurent, A. *Ann. Chim. Phys.* **1836**, *63*, 377. (d) Laurent, A. *Ann. Chim. Phys.* **1837**, *66*, 326. (e) Laurent, A. *Thèse de Docteur*; Paris, 1837, pp 11, 88, 102. (f) Laurent, A. *Comptes Rend.* **1840**, *10*, 179. (g) Laurent, A. *Méthode de Chimie (avec un avis au Lecteur par J. B. Biot)*; Mallet-Bachelier: Paris, 1854.

18. (a) Dumas, J. B. A. *Ann. Chim. Phys.* [2] **1834**, *56*, 113. (b) Dumas, J. B. A. *Ann. Chim. Phys.* [2] **1834**, *56*, 143. (c) Dumas, J. *Ann. Chem. Pharm.* **1839**, *32*, 101. (d) Dumas, J. B. A. *Comptes Rend.* **1839**, *8*, 609. (e) Dumas, J. *Ann. Chem. Pharm.* **1840**, *33*, 179. (e) Dumas, J. *Ann. Chem. Pharm.* **1840**, *33*, 259–300. (f) Dumas, J. *Ann. Chem. Pharm.* **1840**, *35*, 129. (g) Dumas, J.; Peligot, E. *Ann. Chem. Pharm.* **1840**, *35*, 281. (h) Dumas, J.; Piria, R. *Ann. Chem. Pharm.* **1842**, *44*, 66.

19. (a) Gerhardt, C. *Ann. Chim. Phys.* [2] **1839**, *72*, 184. (b) Gerhardt, C. *Comptes Rend.* **1839**, *20*, 1031.

20. Dumas, J. B. A. *Comptes Rend.* **1840**, *10*, 149.

21. (a) Dumas, J. B. A. *Ann. Chim.* **1833**, *54*, 225. (b) Dumas, J. B. A. *Ann. Chim.* **1834**, *56*, 113.

22. (a) Dumas. J. B. A. *Ann. Chim.* **1834**, *56*, 113. (b) Dumas, J. B. A. *Ann. Chim.* **1835**, *57*, 69. (c) Dumas, J. B. A. *Ann. Chim.* **1835**, *57*, 594.

23. Berzelius, J. J. *Versuch über die Theorie der chemischen Proportionen und über die chemische Wirkungen der Eliktrizitat*; Arnold: Dresden, 1820. A Swedish version of this book appeared in 1814 and a French translation of the original Swedish version in 1819.

24. Windler, S. C. H. (Wöhler, F.) *Ann. Chem. Pharm.* **1840**, *33*, 308–310. This paper has been translated into English: Friedman, H. B. *J. Chem. Educ.* **1930**, *7*, 633. It is interesting to note that this paper lists no author when one checks the online Table of Contents for *Liebigs Annalen der Chemie und Pharmacie*.

25. (a) Williamson, A. W. *Philos. Mag.* [iii] **1850**, *37*, 350. (b) Williamson, A. W. *Chem. Gaz.* **1851**, *9*, 294. (c) Williamson, A. W. *Chem. Gaz.* **1851**, *9*, 334. (d) Williamson, A. W. *Quart. J. Chem. Soc.* **1852**, *4*, 229. (e) Williamson, A. W. *Quart. J. Chem. Soc.* **1852**, *4*, 350–355.

26. (a) Hofmann, A. W. *Quart. J. Chem. Soc.* **1849**, *1*, 159. (b) Hofmann, A. W. *Quart. J. Chem. Soc.* **1849**, *1*, 269. (c) Hofmann, A. W. *Quart. J. Chem. Soc.* **1849**, *1*, 285. (d) Hofmann, A. W. *Quart. J. Chem. Soc.* **1850**, *2*, 300. (e) Hofmann, A. W. *Quart. J. Chem. Soc.* **1851**, *3*, 231–240. (f) Hofmann, A. W. *Philos. Trans.* **1850**, *140*, 93. Most of these papers were translated into German: (f) Hofmann, A. W. *Ann. Chem. Pharm.* **1848**, *66*, 129. (g) Hofmann, A. W. *Ann. Chem. Pharm.* **1848**, *67*, 61. (h) Hofmann, A. W. *Ann. Chem. Pharm.* **1848**, *67*, 129. (i) Hofmann, A. W. v. *Ann. Chem. Pharm.* **1849**, *70*, 129. (j) Hofmann, A. W. *Ann. Chem. Pharm.* **1850**, *73*, 91. (k) Hofmann, A. W. *Ann. Chem. Pharm.* **1850**, *73*, 180. (l) Hofmann, A. W. *Ann. Chem. Pharm.* **1850**, *74*, 1. (m) Hofmann, A. W. *Ann. Chem. Pharm.* **1850**, *74*, 117. (n) Hofmann, A. W. *Ann. Chem. Pharm.* **1850**, *75*, 356–368.

27. Gerhardt, C. *Ann. Chim.* **1853**, *37*, 285.

28. Kekulé, F. A. *Ann. Chem. Pharm.* **1857**, *104*, 129.

29. *Classics in the Theory of Chemical Combination*; Benfey, O. T., Translator/ Ed.; Dover: New York, 1963; p 109.

30. Couper, A. S. *Comptes. Rend.* **1857**, *45*, 230.

31. (a) Couper, A. S. *Comptes. Rend.* **1858**, *46*, 1107. This paper was also published in English, Couper, A. S. *Edinburgh. New Philos. J.* (N.S.) **1858**, *8*, 213 and translated into German, Couper, A. S. *Ann. Chem. Pharm.* **1859**, *109*, 369.

32. Ladenburg, A. to Anschütz, R. Letter dated May 12, 1906. The English translation of part of this letter is cited in Dobbin, L. *J. Chem. Educ.* **1934**, *11*, 331: "Couper worked with Wurtz in Paris and asked him to pass on to the Academy his paper on the quadrivalence of carbon. Wurtz, who at the time was not a member of the Academy, was obliged to give the paper to some one else who was a member (usually Balard). He bungled this a little and so Kekulé's communication appeared before Couper's was laid before the Academy. On account of this, great wrath of Couper, who took Wurtz to task and became insolent. This displeased Wurtz and he expelled him from the laboratory. Couper seems to have taken this very much to heart and it was believed in Paris that the beginning of his illness dated from this episode. The story itself is authentic: I have it from Wurtz."

33. For an account of Anschütz' efforts to trace Couper, see Dobbin, L. The Couper Quest. *J. Chem. Educ.* **1934**, *11*, 331.

34. Couper, A. S. *Philos. Mag.* [4] **1858**, *16*, 104.

35. Couper, A. S. *Ann. Chim. Phys.* [3] **1858**, *53*, 472.

36. (a) Kekule, A. *Ann. Chem. Pharm.* **1858**, *105*, 286. (b) Kekule, A. *Ann. Chem. Pharm.* **1861**, *117*, 145. (c) Kekule, A. *Bull. Acad. R. Belg.* [2] **1860**, *10*, 337.

37. Kolbe, H.; Lautemann, E. *Ann. Chem. Pharm.* **1860**, *115*, 157.

38. (a) Anschütz, R. *Ann. Chem. Pharm.* **1885**, *228*, 308. (b) Anschütz, R. *Justus Liebigs Ann. Chem.* **1906**, *346*, 286.

39. (a) Pinkus, A. G.; Waldrep, P. G.; Collier, W. J. *J. Org. Chem.* **1961**, *26*, 682. (b) Pinkus, A. G.; Waldrep, P. G. *J. Org. Chem.* **1966**, *31*, 575.

40. Kekulé, A. *Comptes Rend.* **1859**, *47*, 378–380.

41. Wurtz, A. *Rép. Chim. Pure* **1858**, *1*, 49.

42. Benfey, O. T. From Vital Force to Structural Formulas. In *Classic Researches in Organic Chemistry*; Hart, H., Ed.; Houghton Mifflin: Boston, 1964; p 98.

43. Couper's paper in the *Annales de Chimie et de Physique* had been published while Wurtz was editor of that journal; this author, at least, finds it difficult to conceive that Wurtz simply forgot Couper's contributions, especially in the light of his later statement about Couper's independent development of the basic ideas of structural theory (see ref 44).

44. Wurtz, A. *Dictionnaire de Chimie Pure et Appliquée. Discours Préliminaire. Histoire des Doctrines Chimiques Depuis Lavoisier Jusqu'à Nos Jours*; Librairie Hachette et Cie.: Paris, 1873. Footnote on page lxxi: "It is fair to remember that M. Couper developed analogous ideas without having knowledge of the proposals that had been set out by M. Kekulé, that have had such a large influence on the recent development of organic chemistry."

45. Butlerow, A. *Ann. Chem. Pharm.* **1859**, *110*, 51.

46. For a discussion, see Rocke, A. J. *Brit. J. Hist. Sci.* **1981**, *14*, 27.

47. Butlerov, A. M. Dvenie Babochki Volgo-Uralskoi Fauny (Diurnal Butterflies of the Volga-Ural Fauna. Ph.D. Thesis, Kazan' University, 1849.

48. Butlerov, A. M. *Introduction to the Study of Organic Chemistry*; Kazan', 1861. This book was translated into German: Butlerow, A. *Lehrbuch der organischen Chemie zur Einführung in das specielle Studien derselben*; Leipzig, 1867.

49. (a) Arbuzov, A. E. *Zh. Obshch. Khim.* **1961**, *31*, 2797. Arbuzov, A. E. *J. Gen. Chem. U.S.S.R.* **1961**, *31*, 2611. (b) Bykov, G. V. *J. Chem. Educ.* **1962**, *39*, 220. (c) Bykov, G. V. *Proc. Chem. Soc.* **1960**, 210. (d) Bykov, G. V. *Istoria Klassicheskoi Teory Khimicheskogo Stroenia (History of the Classical Theory of Chemical Structure)*; Akad. Nauk S.S.S.R.: Moscow, 1960. See also (e) Leicester, H. M. *J. Chem. Educ.* **1940**, *17*, 203. (f) Larder, D. F.; Kluge, F. F. *J. Chem. Educ.* **1971**, *48*, 287. (g) Giua, M. *Storia della chimica dall'alchemica alle dottrine moderne*; Torino, Italy, 1946; pp 181, 264. (h) Leicester, H. M. *Adv. Chem. Ser.* **1966**, *61*, 13. (i) Kuznetsov, V. I.; Shamin, A. N. In *The Kekulé Riddle. A Challenge for Chemists and Psychologists*; Wotiz, J. H., Ed.; Cache River Press: Clearwater, FL, 1993; p 211.

50. Staudinger, H.; Meyer, J. *Helv. Chim. Acta* **1919**, *2*, 635.

51. Wittig, G.; Geissler, G. *Liebigs Ann. Chem.* **1953**, *580*, 44.

52. Butlerow, A. *Z. Chem.* **1861**, 549–560. A full English translation of this paper is available: Kluge, F. F.; Larder, D. F. *J. Chem. Educ.* **1971**, *48*, 289–291.

53. See (a) Butlerow, A. *Zeit. Chem.* **1862**, *5*, 297. (b) Butlerow, A. *Zeit. Chem.* **1863**, *6*, 500. (c) Butlerow, A. *Zeit. Chem.* **1864**, *7*, 513.

54. Bulterow, A. *Ann. Chem. Pharm.* **1877**, *189*, 46.

55. (a) Butlerow, A. *Jahresb.* **1863**, 475. (b) Butlerow, A. *Jahresb.* **1864**, 496. (c) Butlerow, A. *Ann. Chem. Pharm.* **1867**, *144*, 132.

56. (a) Boutlerow, A. *Bull. Soc. Chim. France* **1866**, *5*, 30. (b) Butlerow, A. *Z. Chem.* **1870**, *6*, 236. (c) Butlerow, A. *Chem. Zentr.* **1871**, *42*, 89. This paper was abstracted in English: Butlerow, A. *J. Chem. Soc.*, **1871**, *24*, 214. See also Baker, A. A., Jr. *Adv. Chem. Ser.* **1966**, *61*, 81.

57. Horn, D. B. *A Short History of the University of Edinburgh*; University Press: Edinburgh, 1967; p 194.

58. (a) Crum Brown, A. On the Theory of Chemical Combination: A Thesis. M.D. Thesis, University of Edinburgh, 1861. (b) Crum Brown, A. *Trans. R. Soc. Edinburgh* **1864**, *23*, 707. This paper was reprinted in full in Crum Brown, A. *J. Chem. Soc.* **1865**, *18*, 230.

59. Crum Brown, A. *Trans. R. Soc. Edinburgh* **1867**, *24*, 331–339.

60. Crum Brown's obituary, written by "J.W." (James Walker?) appears in *J. Chem. Soc.* **1923**, *123*, 3422.

61. Loschmidt, J. *Sitzungs. kaiser. Akad. Wiss. Wien* **1865**, *52*, 395–413.

62. Loschmidt, J. *Chemische Studien. A. Constitutions-Formeln der organischen Chemie in graphischer Darstellung. B. Das Mariotte'sche Gesetz*; Vienna, 1861. Part A of this work was re-published as Loschmidt, J. *Konstitutions-Formeln der organischen Chemie in graphischer Darstellung. Ostwalds Klassiker der exacten Wissenschaften*, No. 190; Wilhelm Engelmann: Leipzig, 1913.

63. Wiswesser, W. J. *Aldrichimic Acta* **1989**, *22*, 17–19.

64. For an excellent account of Kolbe's career and changing viewpoints, see Rocke, A. J. *The Quiet Revolution. Hermann Kolbe and the Science of Organic Chemistry*; University of California Press: Berkeley, 1993.

65. See (a) Pasteur, L. *Researches on the Molecular Asymmetry of Natural Organic Products*, Alembic Club Reprints; University of Chicago Press; Chicago, 1902. (b) Pasteur, L. *Oevres de Pasteur*; Masson et Cie.: Paris, 1922; Vol. 1, pp 315, 329.

66. Wislicenus, J. *Liebigs Ann. Chem.* **1873**, *167*, 302.

67. Butlerow, A. *Z. Chem. Pharm.* **1862**, *5*, 297.

68. Kekulé, A. *Z. Chem.* **1867**, *3*, 217.

69. (a) See Gillis, J. *Mededelingen van de koninklijke Vlaamse Academie voor Wetenschappen* **1958**, *20*, 3. (b) *Classics in the Theory of Chemical Combination*; Benfey, O. T., Ed.; Dover: New York, 1963; p 1.

70. (a) van't Hoff, J. H. *Voorstel tot Uitbreiding der tegenwoordig in die scheikunde gebruikte Struktuur-Formels in die ruimte; benevens een daarmeé samenhangende opmerkung omtrent het verband tusschen optisch actief Vermogen en Chemische Constitutie van Organische Verbindingen*; Greven: Utrecht, 1874. (b) van't Hoff, J. A. *Arch. Nerland. Sci. Exactes Natur.* **1874**, *9*, 445. This paper was subsequently translated into French, German, and English: (c) van't Hoff, J. H. *Bull. Soc. Chim. France* **1875**, [2] *23*, 295. (d) van't Hoff, J. A. *La chimie dans l'espace*; Bazendijk: Rotterdam, 1875. (e) van't Hoff, J. A. *Die Lagerung der Atome im Raume*; Hermann, F., Translator; Vieweg: Braunschweig, 1877. (f) Marsh, J. E., Translator; *Chemistry in Space*; Clarendon Press: Oxford, 1891.

71. Le Bel, J. A. *Bull. Soc. Chim. France* **1874**, *22*, 337–347. This paper has been translated into English: Richardson, G. M. In *Classics in the Theory of Chemical Combination*; Benfey, O. T., Ed.; Dover: New York, 1963; p 161.

72. (a) Kolbe, H. *J. Prakt. Chem.* [2] **1877**, *15*, 473. An English translation of this diatribe is found in Wheland, G. W. *Advanced Organic Chemistry*, 2nd ed.; John Wiley & Sons: New York, 1949; p 132. A second paper appeared four years later: (b) Kolbe, H. *J. Prakt. Chem.* [2] **1881**, *24*, 405.

73. Walker, J. *J. Chem. Soc.* **1923**, *123*, 940.

74. Glaser, C. *Z. Chem.* **1868**, *4*, 338.

75. (a) Erlenmeyer, E. *Z. Chem.* **1863**, *6*, 728. (b) Erlenmeyer, E. *Ann. Chem. Pharm.* **1866**, *137*, 327.

76. For an interesting history of the rise and subsequent development of stereochemistry, see Ramberg, P. J. *Chemical Structure, Spatial Arrangement. The Early History of Stereochemistry 1874-1914*; Ashgate Publishing Company: Aldershot, U.K., 2003.

Chapter 5

The Atomic Debates Revisited

William H. Brock*

Department of History, University of Leicester, Leicester LE1 7RH, UK
***william.brock@btinternet.com**

Triggered by Benjamin Collins Brodie's extraordinary "Calculus of Chemical Operations" in 1866, the following decade witnessed a number of highly-charged debates about whether chemists should believe in the atomic-molecular theory. Although largely resolved by the complexity of Brodie's system and by its failure to explain stereochemistry, debate was renewed by Ostwald's development of energetics in the 1890s. If energy, not matter, was primary, what need of atoms and molecules when chemical phenomena were sufficiently explained and predicted by energy exchanges? Many chemists, like Ramsay, were nearly persuaded until, in 1906, Perrin rendered scepticism of interest only to historians.

Despite some early signs that not every chemist was happy with Dalton's atomic theory at a philosophical level, the majority of chemical practitioners during the first half of the nineteenth century were content to accept atomism at a conventional level. In the 1960s, David Knight and I called this the "textbook approach" (*1*, *2*). Subsequently, in his important book on chemical atomism, Alan Rocke clarified the distinction between physical atomism and the textbook tradition of chemical atomism, and showed that even when chemists like William Hyde Wollaston asserted that equivalents were more experimentally-based and less hypothetical than atomic weights, a corpuscular philosophy or, at the very least, an idea of a unique invariant of chemically-indivisible units, lay behind their chemical practice (*3*). Chemists were content to be agnostic and indifferent to the question as to whether physical atoms really existed and agnostic and indifferent to whether (if they did) they were the same as their chemical atoms.

Chemical atomism was relatively uncontroversial, even though it was itself a hypothetical construct based upon assumptions concerning stoichiometric units that might vary between individual chemists. But physical atomism was controversial, and if anyone attacked physical atomism, it was bound sometimes to impugn the legitimacy of the chemical atom, despite the latter's usefulness as an analytical stoichiometric category.

That is what happened in 1866 when the complacent conventionalism of British chemists was shattered when the Professor of Chemistry at Oxford, Sir Benjamin Collins Brodie (1817-80), strongly argued that chemistry had gone off the rails; chemists were disobeying Lavoisier's fundamental premise of the Chemical Revolution; namely, that chemistry had to be based upon facts, that is, experimental evidence, not on speculation (4). Since there was no physical evidence for the existence of atoms, the idea of an atomic weight was an absurdity; it was misleading and spurious. The atomic theory, Brodie observed, had recently led to the absurdity of atomic model kits, a phantasmagoria of multi-coloured billiard balls on sticks. Chemistry students were being dangerously misled. Chemistry had to be based upon observable, experimental data such as Gay-Lussac's law of gaseous volumes and Dulong and Petit's rule concerning specific heats. Chemistry also needed to become more mathematical and, ideally, an axiomatic deductive system. And with that declaration to the Royal Society in 1866, he launched a new and very strange system he called "The Calculus of Chemical Operations" which employed Greek symbols for chemical events (that is, operations or reactions), rather than Jacob Berzelius's symbols for chemical elements or atoms. See Table 1.

The maths involved, a form of Boolian algebra, was beyond the ken of most contemporary chemists, with the notable exception of Alexander Williamson who had studied higher mathematics with Auguste Comte in Paris. Consequently, after publishing the first part of his system in 1866 in the *Philosophical Transactions* of the Royal Society, the Chemical Society (whose President, William Allen Miller, was succeeded by Williamson that year) asked Brodie to give a simple explanatory lecture. This exposition entitled "Ideal Chemistry" received massive publicity in William Crookes's *Chemical News* and John Cargill Brough's rival weekly, *The Laboratory* (5). But it is doubtful that Brodie's audience came away much the wiser, having been daunted by a plethora of Greek symbols and Brodie's use of a glass cube which he defined as a 1000-ccs of empty space at NTP. (Despite his attack on molecular models, Brodie was equally dependent upon a visual aid!) What did strike a chord, however, was that Brodie's symbolic system appeared to classify the known chemical elements into three classes: those with a single symbol like hydrogen, α and carbon, κ; those with two identical symbols which Brodie represented algebraically as squared, such as oxygen, ξ^2; and most excitingly, those with a double symbol such as chlorine, $\alpha\chi^2$. The latter group was exciting since the elements involved were formally analogous to compounds such as hydrogen peroxide, $\alpha\xi^2$. Had Brodie uncovered evidence that elements such as chlorine contained hydrogen? His work on organic peroxides in 1864 had already made him suspect chlorine was not an element (6). Had old William Prout, dismissed experimentally by Jean-Baptiste Stas only a few years before, been right all along? Were the elements compounds of hydrogen? (7)

Since this was the period when chemists like William Allen Miller and Edward Frankland were cooperating with astronomers like William Huggins and Norman Lockyer on spectroscopic surveys of the sun and stars, it was easy to speculate (as Brodie himself did) that some of his symbols that carried no earthly elementary meaning, such as χ, might represent elementary materials present in the sun, where dissociation constantly occurred.

Poor Brodie. His well-meaning attempt to place chemistry on surer empirical and mathematical foundations went off at a tangent concerning the possibility that elements were really complex. That is not to say that his anti-physical atomism stance was totally ignored. Far from it. The matter was hotly debated at the Chemical Society in 1867 and in the chemical press for some 18 months with Williamson pressing the case for the acceptance of physical atoms that bore a direct relationship with the chemical atoms defined by Stanislao Cannizzaro's and August Wilhelm Hofmann's revival of Amedeo Avogadro's approach and which bore some sort of relation to the particles of physicists' kinetic theories. As Alan Rocke notes in his essay in this book, and Williamson made explicit in 1869, physical and chemical atomism was supported by a consilience of inductions, namely stoichiometric laws, Cannizzaro's molecular theory and the doctrine of valence. To this evidence must be added the contemporary physicists' kinetic theory of gases as consisting of hard particles. We must remember, too, that while organic chemistry dominated the research headlines in the 1860s and 1870s, the determination of atomic weight values for each and every element remained a perennial theme of inorganic research activity. Moreover, following the Karlesruhe Conference, relative atomic weights became an international collaborative enterprise involving heated discussions over the choice of H = 1 or O = 16 as the standard. Brodie was totally ignored by this movement.

As August Kekulé and others observed, too, the Calculus was based on initial assumptions that, if altered, produced different results; whereas, chemical atomism, as practised, was serving chemists well and forming a self-consistent system. What was the justification for choosing integral values for the indices of Brodie's prime factors, Kekulé and others wanted to know, unless it was an atomic one? Brodie's justification appealed to Gay-Lussac's law, though even rigorous experimentation had not produced whole numbers. In fact, as Duncan Dallas has shown, if non-integral values had been used, negative weights would have appeared (8).

Kekulé's other sensible point was that we do not necessarily choose the simplest hypothesis in science unless the consequences are also simple; and Brodie's consequences were not simple because they produced compounded elements. George Carey Foster, a chemist turned physicist, also noted how our very notion of a "compound" involves a mental division into separate entities so that, unless one appealed to transmutation, the notion of invariable least bits was an inevitable consequence of chemical composition. Ursula Klein has shown how important Berzelius's system of chemical signs was for the development of organic chemistry; by the same token, Brodie's system of signs would have been inhibiting and less conducive to understanding what was happening in chemical reactions and in reaching a model for teaching purposes (9). As Charles Alder Wright pointed out during a debate in 1872, Berzelian symbols could still be used without a commitment to physical atomism. In the striking words of a German historian, Britta Gors, "chemists learned chemistry with the help of the atomic concept, without however having to take the atom's existence into account" (*10*). Ironically, even an arch anti-atomist like Wilhelm Ostwald recognised this in his own influential textbooks.

Brodie's attempt to abolish the chemical atom, and thereby the physical atom, would not work. Nevertheless, Brodie persisted with his Calculus trying to incorporate carbon compounds into the system, but failing to find a simple way of distinguishing between ordinary isomers (which have identical weights and therefore identical symbols), let alone stereochemical ones. He died in 1880 having failed to persuade chemists that his anti-atomistic stance was a sensible way forward now that organic chemists were able to picture molecular structures using atomic symbolism.

Table 1. Brodie's Calculus of Chemical Operations (1866)

From Boole; $x + y = xy$

If a compound weight of 9 units (water) is made up of two substances A and B whose operations for preparing them are x and y ,then xy can stand as the symbol of a single weight whose operation is $(x + y)$.

Let φ be the symbol for a unit of water; φ_1 that for a unit of hydrogen; and φ_2 for a unit of oxygen.

Let $\phi = \alpha^m \xi^{m_1}$; $\varphi_1 = \alpha$ [NB, unjustified]; and $\phi_2 = \alpha^n \xi^{n_1}$

where α and ξ are simple weights (or prime factors) and m, m_1, n, n_1 are simple integers.

From Gay-Lussac's empirical law of volumes:

$$2\alpha^m \xi^{m_1} = 2\alpha + \alpha^n \xi^{n_1}$$

But since $x + y = xy$

$$(\alpha^m \xi^{m_1})^2 = \alpha^2 \alpha^n \xi^{n_1}$$

Collect terms for α and ξ

$$2m = 2 + n \text{ and } 2m_1 = n_1$$

Since the weight of a litre of hydrogen is 1 gram (the necessary definition for a system of weights) and the weight of a litre of water is 9, the simplest solutions are: if n = 0, m = 1 and m_1 = 1 and n_1 =2

Hence, operational symbols are: water, $\alpha\xi$; oxygen ξ^2; hydrogen α, and the compound relative weights are $w(\alpha) = 1$ and $w(\xi) = 8$

During the decade after Brodie's death, physical chemistry took off and offered an alternative route to the mathematization of chemistry. In the hands of Wilhelm Ostwald, it seemed that energy, and the study of thermodynamics generally, offered a solution to the cause of chemical reactivity in a way that simple atomic-molecular theory never had or did. Ostwald was much influenced by a worldwide sense of weariness with materialism, and a renewal of idealism, that gripped European intelligentsia in the 1890s. A group of physicists, among them Ernst Mach, John Bernhard Stallo and Pierre Duhem began to voice doubts about physical atomism because the kinetic theory did not dovetail with accurate experimentation. (For reasons of space, I ignore Marcellin Berthelot's important role in down-playing atomism in French public education (*11*).) The consilience between chemistry and physics had broken down. Mach, in particular, believed science to be a construct of the human mind and that it was not possible to find independent evidence for the existence of matter. Influenced by the thoughts of Georg Helm in 1887, Ostwald began to deny atomism explicitly. He opted instead for energetics – the laws of thermodynamics – rather than mechanical explanations in chemistry. He argued that energy was more fundamental than matter, which he saw only as another manifestation of energy. It followed that chemical events were best analyzed as a series of energy transactions. The difference between one substance and another, including one element and another, was due to their specific energies (*12*).

The claim that energy could and should replace matter and mechanical theories such as the kinetic theory, was debated heatedly at Lübeck in 1895 during a meeting of the Versammlung Deutsche Naturforscher und Ärtze where Ostwald's arguments were supported by Helm and by the Czech, František Wald, who argued that the phase rule was more observationally sound than atomism. Their opponents, Johannes Wislicenus and Victor Meyer for the chemists, and Ludwig Boltzmann, Max Planck, Walther Nernst and others for the physicists, ridiculed Ostwald's contention. Even Mach, who was no believer in atoms, rejected energetics as an alternative.

European chemists were certainly fully aware of this controversy, as private correspondence and journal articles show. However, the majority of chemists, who were chiefly pursuing organic research, where the atom-molecular theory had proved so helpful, ignored the debate. They were surely right to do so. After all, energetics failed to offer any satisfactory explanation for the periodic law, valence, structural theory in organic chemistry, or the so-called relative atomic weights. Despite this indifference, Ostwald persisted, giving a cogent defence of energetics in the Faraday Lecture he gave to the London Chemical Society in 1904 in the lecture theatre of the Royal Institution. There he argued that stoichiometric laws were all deducible from thermodynamics, hence an "atomic hypothesis was unnecessary" (*13*). There are similarities here with the claim that Wollaston's equivalents were independent of a corpuscular hypothesis. By 1904, of course, Ostwald was well aware of radioactivity and radioactive disintegrations. Indeed, in view of such transformations he suggested that elementary transmutations were possible in theory if energy barriers could be breached since elements were merely "hylotropes" in which co-existing energy phases persisted in reactions. But weren't these just chemical atoms in a different disguise? Ostwald may have wished to deny atoms on philosophical grounds, but as a practising chemist he utilised chemical atoms. It was, after all, Ostwald who standardised the *mole* and who argued for O = 16 as the standard for the comparison of relative atomic weights.

William Ramsay was a close personal friend of Ostwald's and he was certainly one of the few British chemists impressed by the lecture. Perhaps this was one of the reasons he began to dabble in spurious claims for transmutation during the next few years – an extraordinary episode recently anayzed by Morrisson (*14*). The majority of chemists ignored Ostwald as they had ignored Brodie forty years earlier. As with the failure of Brodie's calculus of operations in the 1870s, energetics and the phase rule could not compete with the explanatory power of atomism. The textbook tradition of chemical atoms sufficed. The final irony was that when Jean Perrin used the new physical chemistry to indirectly demonstrate the existence of atoms and molecules from Brownian motion, Ernest Rutherford and Frederick Soddy were demonstrating the dissolution of the atom into constituent parts. The physical atom was no longer atomic. Perrin's work convinced Ostwald that atomism was a sound basis for chemical philosophy and he humbly expressed his conviction in the Preface to the 4th edition of his *Outline of General Chemistry* in 1909.

References

1. Brock, W. H.; Knight, D. M. *Isis* **1965**, *56*, 5–25.
2. *The Atomic Debates*; Brock, W. H., Ed.; University of Leicester Press: Leicester, U.K., 1967.
3. Rocke, A. J. *Chemical Atomism in the Nineteenth Century*; Ohio State University Press: Columbus, OH, 1984.
4. Brodie, B. C. *Philos. Trans. R. Soc. London* **1866**, *156*, 781–859.
5. (a) Brodie, B. C. *Chem. News* **1867**, *15*, 295–305. (b) Brodie, B. C. *Laboratory* **1867**, *1*, 230.
6. Brodie, B. C. *Philos. Trans. R. Soc. London* **1848**, *138*, 147–170.
7. *Prout's Hypothesis*, Alembic Club Reprints, No. 20; Dobbin, L.; Kendall, J., Eds.; The Alembic Club: Edinburgh, U.K., 1932.
8. Dallas, D. M. In *Atomic Debates*; Brock, W. H., Ed.; University of Leicester Press: Leicester, U.K., 1967; pp 31–90.
9. Klein, U. *Experiments, Models, Paper Tools*; Stanford University Press: Stanford, CA, 2003.
10. Görs, B. *Chemischer Atomismus*; ERS Verlag: Berlin, 1999, p 18.
11. Bensaude-Vincent, B. *Ann. Sci.* **1999**, *56*, 81–94.
12. Hiebert, E. N. Ostwald. In *Dictionary of Scientific Biography*; Gillispie, C. C., Ed.; Charles Scribner: New York, 1970–1990; Vol. 15, pp 455–469.
13. Ostwald, W. In *Faraday Lectures 1869-1920*; Gibson, C. S., Greenaway, A. J., Eds.; Chemical Society: London, 1928.
14. Morrisson, M. S. *Modern Alchemy. Occultism and the Emergence of Atomic Theory*; Oxford University Press: Oxford/New York, 2007.

Chapter 6

Atoms Are Divisible

The Pieces Have Pieces

Carmen J. Giunta*

Department of Chemistry and Physics, Le Moyne College,
Syracuse, NY 13214
*giunta@lemoyne.edu

The atom as an ultimate and indivisible particle of matter was a
venerable and a viable scientific notion for many years before
and after Dalton. For example, Newton's speculations about
matter in the Queries at the end of his *Opticks* included Particles
"so very hard, as never to wear or break in pieces; no ordinary
Power being able to divide what God himself made one in the
first Creation" (Newton, I. *Opticks*; London, 1704; Query 31).
This chapter describes ideas and scientific evidence from the
late 19th and early 20th centuries about the contrary notion, the
divisibility of atoms. It is about the notion that the "ultimate"
pieces of matter themselves have pieces. It focuses on the
electron and the nucleus, with a few words about the proton and
neutron as well; it does not treat constituent pieces of nucleons
and more exotic particles.

Speculations on Parts of Atoms

The indestructibility of the atom was not universally accepted by scientists
in the 19th century—even by those who accepted the existence of atoms (2).
In chemistry, Prout's hypotheses that atomic weights were integer multiples of
the weight of hydrogen and, more relevant to this topic, that hydrogen was a
constituent of all other elements, emerged within a very few years of Dalton's
atomic theory. Prout's hypotheses refused to die despite repeatedly failing
experimental tests during the course of the 19th century. Perhaps it is more
accurate to say that Prout's second hypothesis refused to die despite his first
hypothesis repeatedly failing experimental tests (3). During the 19th century,

as atomic weights were determined with ever more precision, it became clear that the atomic weights of several elements did not lie within experimental error of integral values. Chlorine was just the most common and most prominent example. Whatever the atomic weight scale employed—hydrogen = 1, oxygen = 16, or any of the other scales proposed or used in those days before international standards—undeniable examples of non-integral values could be found. And yet speculation that atoms were divisible into units of hydrogen atoms, or later halves or quarters of hydrogen atoms, continued among chemists.

Some physicists also tried to look under the hood of atoms, formulating structural models before any of the constituent particles of atoms had been identified. In the 1860s and 1870s, several investigators pursued the notion that oscillations of electrical particles produced the characteristic emissions of light from atoms and molecules which we call spectra. James Clerk Maxwell implicated vibrations of "parts of the molecule" as the cause of bright-line spectra in incandescent gases. George Johnstone Stoney began pursuing this line of inquiry in 1868 (4). In Stoney's model, as it developed over the next few decades, the electrical particles, which he called electrons, were inseparable from atoms and associated with chemical bonds. Thus, Stoney's atoms were still whole, but they had moving parts.

In the 1870s, Wilhelm Weber proposed that the atoms of chemical elements consisted of a massive, negatively charged nucleus and a lighter, positively charged satellite. In 1894 Joseph Larmor presented a theory of the ether in which stable arrangements of "electrons" constitute atoms. In Larmor's theory, electrons were vortices in the ether, permanent and all alike, that were also universal constituents of matter (5).

In 1892, Hendrik Lorentz proposed an electromagnetic theory in which particles (which he called ions) were the source of electromagnetic fields. In 1895, Lorentz explicitly equated his theoretical ions with electrochemical ions. In Lorentz's theory, these electrical charges within matter were the Hertzian oscillators that produced emitted light. Pieter Zeeman carried out experiments on the effect of magnetic fields on spectral lines in order to determine some parameters in Lorentz's theory. In 1896 Zeeman estimated the value of the charge-to-mass ratio of a Lorentzian "ion" that would be consistent with the line splitting he observed. To obtain the observed magnitude of splitting of spectral lines by magnetic fields, the charge-to-mass ratio of these entities had to be much greater than that of electrochemical ions—about three orders of magnitude greater than even the lightest electrochemical ions (5).

The Electron and the Atom

J. J. Thomson is generally credited with "discovering" the electron in 1897 by characterizing cathode rays. Around the centennial of the paper and presentation most commonly cited as the announcement of the discovery, many historians and philosophers of science asked what the phrase "discover the electron" means. After all, the electron has many properties, and it plays an important part in many phenomena. Mass, charge, and spin are among its important properties. Carrier of fundamental electrical charge and building block of atoms and molecules are just two of its roles. Which of the cluster of concepts we now attach to the term electron are defining? Is the attribution of its discovery to Thomson justified, and if so, to 1897?

What Thomson told an audience at the Royal Institution on April 30, 1897, is that cathode rays are charged particles (which he called corpuscles) that are much lighter than ordinary atoms. In a determination involving measurement of heat generated by cathode ray bombardment, he found a mass-to-charge ratio of 1.6×10^{-7} g emu^{-1} (nearly 3 times current value), and he mentioned the similarity of this ratio to the value Zeeman had found for the "ions" of Lorentz. Thomson's lecture included many demonstrations of cathode ray phenomena, including their deflection by electric and magnetic fields. Thomson explicitly speculated that atoms were composed of these corpuscles, and he invoked Prout by name (6).

Clearly, the charge-to-mass ratio is an important characteristic of the electron, arguably its defining characteristic. Thomson was not the first to arrive at a value of this quantity within an order of magnitude of the one accepted today. Might one of those who preceded Thomson in its determination be more justifiably called the discoverer of the electron? Zeeman, for instance? It can be argued that Zeeman's experimental work was too distant from the actual charged particles, whose existence and properties he inferred through Lorentz's theory.

The same cannot be said of Emil Wiechert, who told the Physical Economical Society of Königsberg on January 7, 1897, that cathode rays are moving particles 2000-4000 times lighter than a hydrogen atom. Wiechert based his statement on experimental bounds on the charge-to-mass ratio obtained from magnetic deflection of cathode rays. Wiechert is better known to the history of science as a geophysicist—or at least he was before the centennial of the discovery of the electron drew near. In fact, the entry on Wiechert in the *Dictionary of Scientific Biography* (1976) makes no mention of his work on cathode rays. The scientific achievements it discusses are limited to the geophysical school he founded at Göttingen, whose faculty he joined in 1897. If Wiechert discovered the electron, few took notice.

Walter Kaufmann also reported the determination of the charge-to-mass ratio of cathode rays (about 10^7 emu g^{-1}) in a paper he submitted in April 1897 (7). Kaufmann also based his result on magnetic deflection measurements; however, he concluded that the hypothesis of cathode rays as emitted *particles* could not explain his data. (One of the outstanding questions in the study of cathode rays in the late 1890s was whether they were particles or electromagnetic waves. Thomson and Kaufmann were typical of their countrymen: most British researchers leaned toward the particulate hypothesis and most Germans toward waves.) Today Kaufmann is better known for his careful measurements of the velocity-dependent mass of the electron published over several years beginning in 1901; these results were later explained by special relativity.

Thomson was awarded the Nobel Prize in physics in 1906 "in recognition of the great merits of his theoretical and experimental investigations on the conduction of electricity by gases." The presentation speech, by J. P. Klason, President of the Royal Swedish Academy, highlights his work in characterizing electrons (8). (Klason called them electrons; Thomson, in his award lecture, still called them corpuscles.) Kaufmann's work was not similarly recognized; however, he was recommended for the prize (along with Thomson) by Marie and Pierre Curie (9).

In August 1897, Thomson submitted a paper to *Philosophical Magazine* which included accounts of two independent methods of mass-to-charge measurements, arriving at about 10^{-7} g emu^{-1} (less than twice today's accepted value) independent of the gas in the discharge tube or the metal of the cathode. This paper included some speculation of how the corpuscles in atoms might arrange themselves in concentric shells which somehow mimic chemical periodicity (10). Thus, Thomson not only characterized cathode rays, a rather arcane and artificial physical phenomenon; he placed those cathode-ray "corpuscles" into a much more general and fundamental context as key entities in the structure of matter.

Isobel Falconer (11) and later Helge Kragh (12) have documented how Thomson's thinking in the 1880s and 1890s prepared him to thrust the cathode-ray corpuscles into the middle of a theory of matter. During most of that time, Thomson was not interested in cathode rays, but in dynamical theories of matter. As far back as 1882, Thomson mathematically worked out equilibrium configurations of systems of vortices, which might somehow comprise atoms. The notion that atoms might be vortices in the ether had first been proposed by William Thomson (who was later to become Baron Kelvin) in 1867. J. J. Thomson worked out stable arrangements of 2, 3, 4, 5, and 6 vortices. He also described experiments Alfred Mayer had conducted with floating magnets as a useful analog for larger numbers of vortices, whose mathematical treatment would be messy if not intractable. In 1892, he pointed out analogies between Mayer's magnets, arrangements of vortices, and chemical periodicity. In an 1896 lecture, he spoke of the likelihood that chemical elements were composed of a primordial element, as Prout had suggested. Philipp Lenard's 1896 work on cathode rays led Thomson to think that that prime matter was electrical, and it was late in 1896 that he turned in earnest to characterizing those rays.

It has been suggested that Thomson deserves credit for discovering the electron not in 1897 but in 1899. In that year, he reported not just the mass-to-charge ratio, but the charge separately (and therefore the mass as well), thereby providing convincing proof that the electron was much lighter than the lightest known atom. He reported the charge of the corpuscle to be 6.8×10^{-10} esu (about 1.4 times the value currently accepted) using the cloud chamber technique developed by his student C. T. R. Wilson. In the same paper, he identified photoelectric and thermoelectric particles as the same as his corpuscles. And he proposed a model of the atom (essentially, a qualitative version of the plum-pudding model) and a mechanism for ionization involving detachment of corpuscles (13). By 1899, then, Thomson had accumulated considerable evidence for what he had proposed in 1897, namely that corpuscles were much lighter than even the lightest atom, and that they were much more widespread than cathode ray tubes.

Clearly there is something arbitrary about defining the essence of discovery and attributing it to a particular researcher at a particular time. In my mind, the identity of the electron as a building block of ordinary matter is a key part of the concept of the electron; simply characterizing cathode rays as particles of a certain charge-to-mass ratio would have been a far more limited discovery (albeit perhaps better defined). The question of timing is also somewhat arbitrary. Clearly speculation is not the same as discovery; however, on the other hand, discovery is not the same as proof beyond a shadow of a doubt. In my opinion, Thomson had more than speculation about the role of corpuscles in 1897, although he had a good deal more evidence by 1899.

Thomson's model of atomic structure was based on only one kind of subatomic particle, the only one yet known, his "corpuscles" (our electrons). Because the mass of the electron was known at least approximately, even the simplest atoms had to have thousands of electrons. And because of the negative charge of the electron, the space occupied by those electrons had to be positively charged. Thomson's mathematically formidable 1904 paper (*14*) on his atomic model tested the question of the possible stability of his atom against loss of energy by electromagnetic radiation (an issue mentioned more frequently in connection with the later Bohr model of the atom) (*12*). Within a couple of years, though, Thomson concluded that the number of electrons in atoms was only the order of magnitude of the gram atomic weight (*15*). By implication, then, the atom had to have *other* pieces which accounted for most of its mass.

Radioactivity and Atomic Fragments

Radioactivity is the window through which the first glimpses of the more massive pieces of the atom were obtained. Radioactivity proved to be a remarkably fruitful field for science in very short order. Less than a decade after Henri Becquerel's first reports of penetrating rays from uranium salts (*16*), Ernest Rutherford and Frederick Soddy concluded that radioactivity was a subatomic phenomenon. A paragraph near the end of their landmark 1902 paper contains their conclusions: "Since, therefore, radioactivity is at once an atomic phenomenon and accompanied by chemical changes in which new types of matter are produced, these changes must be occurring within the atom, and the radioactive elements must be undergoing spontaneous transformation. ... Radioactivity may therefore be considered as a manifestation of subatomic chemical change (*17*)."

The idea of one element changing into another was foreign to the chemistry and physics of the time because of its association with discredited, pre-scientific notions. Rutherford is supposed to have urged his colleague, "Don't call it transmutation. They'll have our heads off as alchemists (*18*)." Whatever they called it, they saw that atoms can change their identity. They also saw that these changes could be used to shed further light on the inner workings of atoms. In this same paper, they expressed the hope that radioactivity would serve as a tool for "obtaining information of the processes occurring within the chemical atom" (*17*).

In 1902, Rutherford was a young globe-trotting academic. Born and educated in New Zealand, he won a scholarship that brought him to J. J. Thomson's laboratory in Cambridge. While there, he distinguished two kinds of radiation from uranium characterized by their different powers of penetrating matter. We still employ the unimaginative labels he used for convenience, α and β (*19*). Rutherford moved to North America to take a faculty position at McGill University in Montreal. There he discovered an isotope of radon, as his radioactive emanation from thorium would come to be known (*20*), and he worked out the time-dependence of radioactive emission, introducing the term half-life (*21*). Soddy arrived in Montreal in 1900 to take the post of Demonstrator of chemistry at McGill. Soddy, unlike Rutherford, was actually a chemist, having earned first class honors in that subject at Oxford. Rutherford's only credential in chemistry was the Nobel Prize he would receive in 1908.

Nature of the α Particle

Ever since Rutherford had distinguished α from β rays, he was engaged in investigating the nature of α particles and their interactions with matter. Over the course of about a decade, he came to the conclusion that α particles were doubly-charged helium atoms. (We would say helium nuclei, but he did not know about nuclei—yet.) The nature of the α particle was the subject of Rutherford's Nobel lecture in 1908 (*22*). The last piece of evidence for the identity of the α particle as a doubly-charged helium ion came from an experiment carried out by Rutherford and his research student Thomas Royds. Before the elegant experiment described in that paper, work by Rutherford, his associates, and independent investigators had measured the charge-to-mass ratio of the α particle and had noted the appearance of helium in the presence of α emitters at the same rate as the emission of α particles. What Rutherford and Royds did was to separate the collection of helium from the α-emitting source by something that functioned as a semi-permeable membrane, penetrable by α particles but not by ordinary gases (*23*).

Scattering α Particles and the Nucleus

Even before he finished characterizing α particles, Rutherford had begun using them as a probe of the atom. The discovery of the nucleus consisted of the realization that all of an atom's positive electrical charge and most of its mass are confined to a volume that is only a tiny fraction of the atom's volume. Rutherford published these conclusions in 1911 (*24*), and he reached them on the basis of experiments involving the scattering of α particles. Discovery of the nucleus is the main reason for Rutherford's inclusion in introductory chemistry textbooks. Diagrams purporting to depict Rutherford's α-particle scattering experiments often accompany the discussion of the discovery of the nucleus in introductory chemistry texts. Such diagrams are conceptually clear but a bit misleading historically; they depict not the experiments that led Rutherford to discover the nucleus but experiments that confirmed the discovery.

In 1906, Rutherford published a paper titled "Retardation of the α Particle from Radium in passing through Matter (*25*)." The apparatus is shown in Figure 1 as a schematic from the paper and as a line drawing from *The Restless Atom* (*26*). α particles were made to pass through an evacuated brass tube through a slit to be detected by a photographic plate. The slit could be covered by thin layers of matter such as aluminum foil or mica. Rutherford mentioned scattering of α rays only briefly near the end of this paper, as a complicating factor in the retardation studies.

He promised further experiments to clear up the matter, and Hans Geiger was the man who carried them out in Rutherford's lab. (That lab was by now back across the Atlantic, at the University of Manchester.) Figure 2 shows Geiger's apparatus, which was physically different but conceptually similar to the earlier one: a tube containing a source of α particles at one end, a screen with a slit toward the middle, and a detector at the other end (*27*).

Photographic plates from Rutherford's 1906 paper appear in Figure 3. They show the image made by the α particles from the radioactive wire as they passed through the slit. The blurrier lower half of the photo on the right was produced by α particles that also passed through a thin piece of mica, which covered the lower half of the slit. Some of the rays were deflected by as much as 2° in the course of passing through the 30-μm thick mica, requiring, Rutherford estimated, an average electric field of 100 million volts per cm, "a deduction in harmony with the electronic theory of matter." Geiger's 1908 results appear in Figure 4. Curve A represents the distribution of scintillations when the α particles passed through an evacuated tube. Curve B represents the distribution when one piece of gold leaf covered the slit, and curve C two pieces. "It will be noticed," Geiger wrote, "that some of the α-particles after passing through the very thin leaves were deflected through quite an appreciable angle." Further experiments were promised, and Geiger published an extensive account of the small-angle scattering in 1910 (*28*).

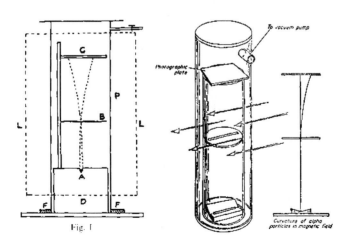

Figure 1. Rutherford's apparatus for studying the retardation of α particles as depicted in his 1906 paper (25) (left) and in reference (26) (right).

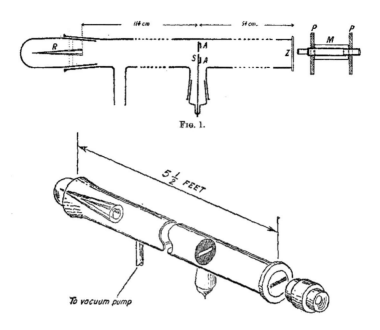

Figure 2. Hans Geiger's apparatus for studying the scattering of α particles as depicted in his 1908 paper (27) (top) and in reference (26) (bottom).

Fig. 3B

Figure 3. Photographs from Rutherford's 1906 paper (25) on retardation of α particles compare particles shot straight through a slit (left) to those passed through a thin layer of mica (right).

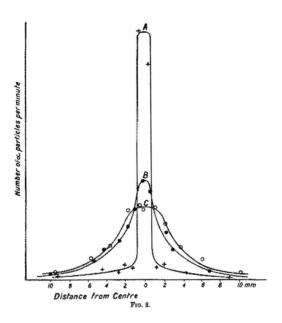

Distance from Centre
Fɪɢ. 2.

Figure 4. Plot from reference (27) of the distribution of deflections of particles that have passed through zero, one, and two sheets of gold leaf.

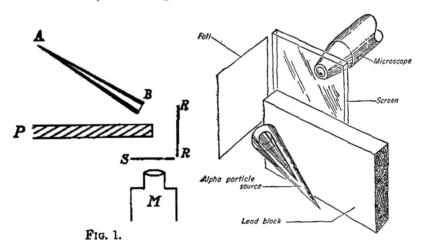

Fɪɢ. 1.

Figure 5. Geiger and Marsden's apparatus for large-angle scattering of α particles as depicted in their 1909 paper (29) (left) and in reference (26) (right).

Geiger also looked into large-angle scattering in Rutherford's lab, working now with Ernest Marsden. If α particles could be *reflected* from matter, the simple apparatus depicted in Figure 5 would be certain to detect it. An α-particle source was placed near a metal plate or foil, and a detector placed nearby so as to intercept any reflected α-particles. A lead plate was placed to block any α-particles that might otherwise pass directly from the source to the detector. Sure enough, they found that metal plates were able to reflect rather than absorb some particles. But

73

even a single piece of gold foil less than a micron thick could deflect some particles through 90° (*29*). "It was almost as incredible as if you fired a fifteen-inch shell at a piece of tissue paper and it came back and hit you," Rutherford recalled many years later (*30*).

Rutherford's 1911 paper (*24*) was based mainly on these experimental investigations of scattering. Titled "The Scattering of α and β Particles by Matter and the Structure of the Atom," it was a heavy-duty mathematical analysis of the sort Rutherford was not known for. He developed the consequences of scattering from a center carrying a charge of about 100 times the electron charge (positive *or negative*) surrounded by a sphere containing an equal and opposite charge spread uniformly throughout it. Of course the mathematical theory was amenable to specific predictions, and Rutherford mentioned at the end of his paper that Geiger and Marsden were already beginning to test the theory. The results of those investigations were published in 1913 (*31*). Here we finally have an apparatus like the one depicted in textbooks, capable of measuring a large range of scattering angles (shown in Figure 6). It included a rotatable circular stage on which a microscope and zinc sulfide detection screen were mounted. The α source was largely surrounded by lead so as to direct emerging particles to the scattering target.

Pieces of the Nucleus

Simplicity was never an attribute attached to the concept of nucleus in the way that it was attached for many years to the concept of atom. By the time the nuclear model of the atom was formulated, the atom was already known to be composite and at least some atoms impermanent. If the nucleus contained most of the mass of an atom, and if a radioactive atom spontaneously ejected pieces like α particles, then the nucleus must be where those ejected particles came from. Of course, we teach our students that nuclei are made of protons and neutrons. How did these particles come to be known? Rutherford had a hand in their stories as well.

The Proton

In 1914, Rutherford noted "It is well known from the experiments of J.J. Thomson and others, that no positively charged carrier has been observed of mass less than that of the hydrogen atom. The exceedingly small dimensions found for the hydrogen nucleus [based on scattering of α particles by hydrogen] add weight to the suggestion that the hydrogen nucleus is the positive electron ..." By positive electron, Rutherford meant fundamental particle of positive electricity or something similar. He went on to speculate, "The helium nucleus has a mass nearly four times that of hydrogen. If one supposes that the positive electron, i.e. the hydrogen atom, is a unit of which all atoms are composed, it is to be anticipated that the helium atom [that is, the helium nucleus] contains four positive electrons and two negative (*32*)." (The notion of negative electrons in the nucleus was natural. After all, β particles had already been identified as electrons, and the emission of β particles was different than ordinary ionization. It was radioactivity, a cluster of phenomena with which the nucleus was associated as soon as the concept of nucleus was introduced.)

Barrie Peake (*33*) dates the discovery of the proton to 1913, when Rutherford concluded that hydrogen has the "simplest possible structure of a nucleus with one unit charge (*34*)." This paper of Rutherford's does not, however, contain the notion of hydrogen nucleus as atomic building block, something that I consider crucial to the concept of the proton. Indeed, one can say that the ionized hydrogen atom was known long before 1913 in discharge tubes and (in solvated form) in electrochemistry; however, there was no good reason to think then that ionized hydrogen was a building block of atoms.

For years, these particles had no special name. Rutherford suggested two names for the hydrogen nucleus at the 1920 British Association meeting in Cardiff. Rutherford spoke on "The Building up of Atoms," and he speculated on the constitution of several composite bodies as combinations of hydrogen nuclei and electrons. (One of the composite particles he did mention by name was the yet "unknown doublet neutron of mass 1 and charge 0" composed of a hydrogen nucleus and an electron "in very close juxtaposition;" more on that below.) Oliver Lodge asked Rutherford for a term for the hydrogen nucleus so as to avoid confusion with the hydrogen atom, and Rutherford offered two: "prouton" or "proton (*35*)." The former term would honor William Prout. The later can be derived from the Greek "first" or "primary."

By this time, Rutherford had evidence that hydrogen nuclei could come from more complex nuclei (*36*). In 1919, he reported an anomalous effect when he subjected dry air to α particles: H atoms seemed to be produced even when there was no hydrogen in the system! Rutherford correctly interpreted the presence of H nuclei as a sign that the α particles caused some sort of transmutation, and that the H nuclei were fragments of that reaction. I infer from Rutherford's use of the term "disintegration" that he pictured the reaction as an induced fission of nitrogen. In fact, the reaction was $^{14}N + ^4He \rightarrow ^{17}O + ^1H$. Rutherford and his associates worked to bolster the conclusion of the proton as a fragment of a nuclear reaction over the next five years or so by more rigorously excluding hydrogen impurities and by finding proton fragments in collisions of α particles with a variety of light elements (*35*).

The Neutron

Unlike the proton, which was observed long before it was named, the neutron was named before it was observed. Rutherford's talk at the British Association meeting in Cardiff in late summer 1920 seems to be the first usage of the term neutron with a meaning similar to our understanding of the term. Rutherford noted that the composite neutron, a "close association" of a hydrogen nucleus and electron, would have interesting properties. Except at very short distances, it would give rise to a negligible electric field, which would allow it to pass easily through matter, and he doubted that it could be confined to a material container. It could, however, have a place within the nucleus, kept there by the immense electric field within the tiny space of the nucleus. A few weeks earlier, in his Bakerian Lecture before the Royal Society, Rutherford (who by now had succeeded J. J. Thomson as the Cavendish Professor at Cambridge) described this composite particle without naming it. He also announced his intent to search for such a particle in electrical discharges of hydrogen, where hydrogen nuclei and electrons would be present in a high-energy environment (37).

Several students and associates of Rutherford attempted to make and detect neutrons by these means or via "swift" protons produced in nuclear reactions induced by α particles: protons and electrons were given opportunities to smash into each other. These efforts did not lead to the discovery of the neutron.

The key event that eventually *did* lead to its discovery was a report by Walther Bothe and Herbert Becker at Giessen, in an experiment whose purpose had nothing to do with neutrons. Bothe and several collaborators were investigating transformations of various light nuclei when exposed to α particles. In 1930, Bothe and Becker reported a very penetrating radiation from beryllium exposed to α particles. The radiation was more penetrating than the most penetrating γ rays known, and the investigators thought that they were most likely γ rays (38).

Irene and Frédéric Joliot-Curie investigated this system further, communicating three reports on the putative γ rays in 1931 and early 1932. Their report from January 1932 described the effect of interposing various materials between the beryllium source of the rays and the ionization chamber they used as a detector. Thin screens of several metals had little or no effect on the emissions. But when hydrogen-containing materials were interposed, the ionization current increased. The augmentation was most pronounced with paraffin. The Joliot-Curies realized that the increased ionization was due to protons coming from the screen: the radiation from the beryllium could knock protons from paraffin and other hydrogen-containing materials (39).

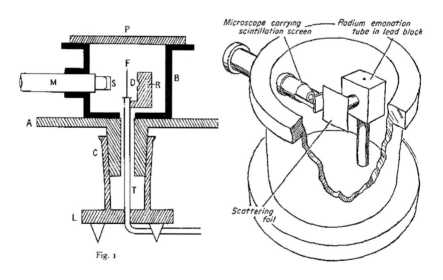

Figure 6. Geiger and Marsden's iconic apparatus for studying the scattering of α particles as depicted in their 1913 paper (31) (left) and in reference (26) (right).

76

These results prompted James Chadwick to look into this experimental system. The ability of the radiation from beryllium to dislodge protons suggested to him that the radiation was in the form of a massive particle rather than a quantum (or photon, as we would say). A particle with mass about equal to that of a proton would make conservation of momentum work out easily in the observed phenomena. Chadwick reproduced the Curie's observation with paraffin, and he added experiments in which he allowed the radiation from beryllium to pass through samples of other light elements. In these cases, he observed short-ranged recoil atoms, again consistent with the beryllium radiation being a material particle. The radiation would have to be electrically neutral, however, to explain its extreme power of penetration. Within a month, Chadwick sent a letter to *Nature* describing his experiments and proposing that they were best explained if the radiation was a material particle carrying no charge and mass about equal to that of a proton—that is, a neutron (*40*). A more detailed paper followed three months after that.

Chadwick was one of Rutherford's associates in looking for neutrons in the 1920s. He had been a student of Rutherford's at Manchester, receiving an M.Sc. in 1913. He went to Berlin to work with Hans Geiger, who had worked with Rutherford for many years at Manchester. As an English civilian in Germany, Chadwick was interned during the Great War. He caught up again with Rutherford at the Cavendish Laboratory in 1919, and Chadwick remained at Cambridge until 1935. That same year, he received the Nobel Prize in physics for discovery of the neutron (*41*).

Is the neutron as we understand it today really a combination of a proton and electron as Rutherford envisioned it? This is a philosophically interesting question. When neutrons decay, they produce a proton and an electron (and an antineutrino as well); however, these particles are not understood to have a real existence "within" an intact neutron. Indeed, the constituent parts of neutrons (and protons for that matter) are understood to be quarks. The phenomenon of neutron decay is explained by a transformation of one of its constituent quarks, turning the neutron into a proton; the energy difference between the neutron and proton gives rise to the electron and antineutrino.

Conclusion

Most of the developments described in this chapter span a period from the 1890s to the early 1930s. At the beginning of that period, skepticism over the very existence of atoms was still respectable among both chemists and physicists. Among those who believed in physical existence of atoms, the default picture was generally one of fundamental solidity and permanence (which is not to deny that some scientists thought about possible constituent parts). By the end of the period, scientists universally accepted both the reality of atoms and their complexity. Not only did some atoms fall apart all by themselves, but human beings could break off pieces. Atoms were no longer the ultimate pieces of matter. By the 1930s, not only was it clear that these pieces had pieces: it was likely that at least some of the pieces of pieces had pieces.

Thus the transformation of the default picture of the fundamental structure of matter underwent a thorough transformation during this time—and we have not even mentioned the quantum revolution during the latter portion of this same time. The foundations of matter must have appeared none too solid to many of the researchers of the period.

Let us close with a revealing bit of doggerel, illustrating how the concept of the atom was changing before chemists' very eyes. The author is eminent British chemist William Ramsay, best known for his role in the discovery and isolation of most of the noble gas elements, for which he was awarded the 1904 Nobel Prize in Chemistry. He wrote these verses for his 1902 lab dinner.

The Death Knell of the Atom

Old time is a-flying; the atoms are dying;
Come, list to their parting oration:-
"We'll soon disappear to a heavenly sphere
"On account of our disintegration,

"Our action's spontaneous in atoms uranious
"Or radious, actinious, or thorious;
"While for others the gleam of a heaven-sent beam
"Must encourage our efforts laborious.

"For many a day we've been slipping away
"While the savants still dozed in their slumbers;
"Till at last came a man with gold-leaf and tin can,
"And detected our infinite numbers."

So the atoms, in turn, we now clearly discern,
Fly to bits with the utmost facility;
They wend on their way, and in splitting, display
An absolute lack of stability.

'Tis clear they should halt on the grave of old Dalton
On their way to celestial spheres;
And a few thousand million - let's say a quadrillion -
Should bedew it with reverent tears.

There's nothing facetious in the way that Lucretius
Imagined the chaos to quiver;
And electrons to blunder, together, asunder,
In building up atoms for ever.

William Ramsay, 1902 (*42*)

Notes and References

Some of the primary sources listed below are available in transcription or page image on the open internet. For these items, URLs have been listed as well as standard bibliographic reference.

1. Newton, I. *Opticks*; London, 1704; Query 31; http://web.lemoyne.edu/~giunta/NEWTON.HTML.
2. Brock, W. H. The Atomic Debates Revisited. In *Atoms in Chemistry: From Dalton's Predecessors to Complex Atoms and Beyond*; Giunta, C. J.; Ed.; ACS Symposium Series 1044; American Chemical Society: Washington, DC, 2010; Chapter 5.
3. Brock, W. H. *From Protyle to Proton: William Prout and the Nature of Matter, 1785–1985*; Adam Hilger, Ltd.: Bristol, U.K./Boston, MA, 1985.
4. Hentschel, K. *Mapping the Spectrum*; Oxford University Press: Oxford, 2002; p 305.
5. Arabatzis, T. Rethinking the "discovery" of the electron. *Studies in the History and Philosophy of Modern Physics* **1996**, *27*, 405–435.
6. Thomson, J. J. Cathode Rays. In *Proceedings of the Royal Institution*; 1897; Vol. *15*, pp 419–432. Available in Bragg, W. L., Porter, G., Eds.; *Royal Institution Library of Science*; Elsevier: Barking, Essex, England, 1970; Vol. 5, pp 36–49.
7. Kaufmann, W. Die magnetische Ablenkbarkeit der Kathodenstrahlen und ihre Abhangigkeit vom Entladungspotential. *Annalen der Physik und Chemie* **1897**, *61*, 544–552.
8. The Nobel Prize in Physics 1906. Nobel Foundation Web site. http://nobelprize.org/nobel_prizes/physics/laureates/1906/.
9. Falconer, I. Corpuscles to Electrons. In *Histories of the Electron: The Birth of Microphysics*; Buchwald, J. Z., Warwick, A., Eds.; MIT Press: Cambridge, MA, 2001; p 98.
10. Thomson, J. J. Cathode rays. *Philos. Mag.* **1897**, *44*, 293–316; http://web.lemoyne.edu/~giunta/thomson1897.html.
11. Falconer, I. Corpuscles, electrons and cathode rays: J. J. Thomson and the "discovery of the electron.". *Br. J. Hist. Sci.* **1987**, *20*, 241–276.
12. Kragh, H. J. J. Thomson, the electron, and atomic architecture. *Phys. Teach.* **1997**, *35*, 328–332.
13. Thomson, J. J. On the masses of the ions in gases at low pressures. *Philos. Mag.* **1899**, *48*, 547–567; http://www.chemteam.info/Chem-History/Thomson-1899.html.
14. Thomson, J. J. On the structure of the atom: An investigation of the stability and periods of oscillation of a number of corpuscles arranged at equal intervals around the circumference of a circle; with application of the results to the theory of atomic structure. *Philos. Mag.* **1904**, *7*, 237–265.
15. Thomson, J. J. On the number of corpuscles in an atom. *Philos. Mag.* **1906**, *11*, 769–781.

16. (a) Becquerel, H. Sur les radiations émises par phosphorescence. *Comptes Rendus* **1896**, *122*, 420−421. (b) Becquerel, H. Sur les radiations invisibles émises par les corps phosphorescents. *Comptes Rendus* **1896**, *122*, 501−503. English translations available at http://web.lemoyne.edu/~giunta/EA/ BECQUERELann.HTML.

17. Rutherford, E.; Soddy, F. The cause and nature of radioactivity. *Philos. Mag.* **1902**, *4*, 370−396; http://web.lemoyne.edu/~giunta/ruthsod.html.

18. Pais, A. *Inward Bound*; Oxford University Press: New York, 1986; p 113.

19. Rutherford, E. Uranium radiation and the electrical conduction produced by it. *Philos. Mag.* **1899**, *47*, 109−163; http://www.chemteam.info/Chem-History/Rutherford-Alpha&Beta.html.

20. Marshall, J. L.; Marshall, V. R. Ernest Rutherford, the "true discoverer" of radon. *Bull. Hist. Chem.* **2003**, *28*, 76−83; http://www.scs.uiuc.edu/ ~mainzv/HIST/awards/OPA%20Papers/2003-Marshall.pdf.

21. Rutherford, E. A radioactive substance emitted from thorium compounds. *Philos. Mag.* **1900**, *49*, 1−14; http://www.chemteam.info/Chem-History/ Rutherford-half-life.html.

22. Rutherford, E. Nobel Lecture: The Chemical Nature of the Alpha Particles from Radioactive Substances, December 11, 1908. Nobel Foundation Web site. http://nobelprize.org/nobel_prizes/chemistry/laureates/1908/ rutherford-lecture.html.

23. Rutherford, E.; Royds, T. The nature of the α particle from radioactive substances. *Philos. Mag.* **1909**, *17*, 281−286; http://web.lemoyne.edu/ ~giunta/EA/ROYDSann.HTML.

24. Rutherford, E. The scattering of α and β particles by matter and the structure of the atom. *Philos. Mag.* **1911**, *21*, 669−688; http://www.chemteam.info/ Chem-History/Rutherford-1911/Rutherford-1911.html.

25. Rutherford, E. Retardation of the α particle from radium in passing through matter. *Philos. Mag.* **1906**, *12*, 134−146.

26. Romer, A. *The Restless Atom*; Doubleday: Garden City, NY, 1960.

27. Geiger, H. On the scattering of the α-particles by matter. *Proc. R. Soc. London, Ser. A* **1908**, *81*, 174−177.

28. Geiger, H. The scattering of the α-particles by matter. *Proc. R. Soc. London, Ser. A* **1910**, *83*, 492−504.

29. Geiger H.; Marsden, E. On a diffuse reflection of the α-particles. *Proc. R. Soc. London, Ser. A* **1909**, *82*, 495−500; http://www.chemteam.info/Chem-History/GM-1909.html.

30. Just when Rutherford first used this colorful and oft-quoted phrase is not clear. It appears in print, however, in Rutherford, E. The Development of the Theory of Atomic Structure. In Needham, J., Pagel, W.; Eds.; *Background to Modern Science*; Cambridge University Press: Cambridge, U.K., 1938. The text is from a lecture Rutherford gave in 1936.

31. Geiger, H.; Marsden, E. The laws of deflexion of α particles through large angles. *Philos. Mag.* **1913**, *25*, 604−623; http://www.chemteam.info/Chem-History/GeigerMarsden-1913/GeigerMarsden-1913.html.

32. Rutherford, E. The structure of the atom. *Philos. Mag.* **1914**, *27*, 488−498; http://www.chemteam.info/Chem-History/Rutherford-1914.html.

33. Peake, B. M. The discovery of the electron, proton, and neutron. *J. Chem. Educ.* **1989**, *66*, 738.

34. Rutherford, E.; Nuttall, J. M. Scattering of α particles by gases. *Philos. Mag.* **1913**, *26*, 702–712.

35. Romer, R. Proton or prouton?: Rutherford and the depths of the atom. *Am. J. Phys.* **1997**, *65*, 707–716.

36. Rutherford, E. Collisions of alpha particles with light atoms. IV. An anomalous effect in nitrogen. *Philos. Mag.* **1919**, *37*, 581–587; http://web.lemoyne.edu/~giunta/rutherford.html.

37. Rutherford, E. Bakerian lecture. Nuclear constitution of atoms. *Proc. R. Soc. London, Ser. A* **1920**, *97*, 374–400.

38. Walther Bothe and the Physics Institute: The Early Years of Nuclear Physics. Nobel Foundation Web site. http://nobelprize.org/nobel_prizes/medicine/articles/states/walther-bothe.html (accessed August 12, 2009).

39. Curie, I.; Joliot, F. Émission de protons de grande vitesse par les substances hydrogénées sous l'influence des rayons γ très pénétrants. *Comptes Rendus* **1932**, *194*, 273–275.

40. Chadwick, J. Possible existence of a neutron. *Nature* **1932**, *192*, 312; http://www.nature.com/physics/looking-back/chadwick/index.html.

41. James Chadwick: Biography, 1965. Nobel Foundation Web site. http://nobelprize.org/nobel_prizes/physics/laureates/1935/chadwick-bio.html (accessed August 12, 2009).

42. Ramsay, W. The death-knell of the atom. *Nature* **1905**, *73*, 132; http://www.chem.ucl.ac.uk/resources/history/chemhistucl/hist16a.html.

Chapter 7

Eyes To See: Physical Evidence for Atoms

Gary Patterson*

Department of Chemistry, Carnegie Mellon University, Pittsburgh, PA 15213
*gp9a@andrew.cmu.edu

A series of episodes in the historical development of our view of chemical atoms are presented. Emphasis is placed on the key observations that drove chemists and physicists to conclude that atoms were real objects and to envision their structure and properties. The kinetic theory of gases and measurements of gas transport yielded good estimates for atomic size. The discovery of the electron, proton and neutron strongly influenced discussion of the constitution of atoms. The observation of a massive, dense nucleus by alpha particle scattering and the measurement of the nuclear charge resulted in an enduring model of the nuclear atom. The role of optical spectroscopy in the development of a theory of electronic structure is presented. The actors in this story were often well rewarded for their efforts to see the atoms.

Introduction

The current paradigm in chemistry celebrates the existence of physical entities called chemical atoms (now known simply as atoms). John Dalton (1766-1844) looked at the material world in which he lived and visualized it in terms of a set of different material objects of small size and combining capacity (*1*). He called these particles atoms in his *New System of Chemical Philosophy (1808)*. Others, such as Humphry Davy (1778-1829), were not yet willing to see the world in this way. Dalton combined both a particular theory of nature and specific observations to arrive at his views. The present paper will examine some episodes in the history of chemistry that enabled other chemists to "see" atoms as appropriate chemical constituents of our world. The view of what constitutes a chemical atom has changed during the time period from 1808 to 2008, but the common theme requires a context in which actual measurements can be viewed as "evidence for atoms."

Sir J.J. Thomson (1856-1940), in his Romanes lecture of 1914 on "The Atomic Theory" stated that "it affords a striking proof that a theory can only grow by the cooperation of thought and facts, and that all that is valuable in a physical theory is not only tested, but in most cases suggested, by the study of physical phenomena (*2*)." He was one of many British scientists who played a vital part in the story of the atom and who were knighted in the process. As the evidence regarding atoms changed, his views on the atom changed as well. He learned to see with new eyes on a regular basis.

The Eyes of Thermodynamics and Kinetic Theory

The 19th century was a period when the properties of matter as a function of temperature and pressure were summarized by thermodynamic descriptions. Many of the key concepts were pioneered by Rudolph Clausius (1822-1888). One of the most important concepts championed by Clausius was that the particles of the gas were in constant motion (1857). The observable pressure of the gas was then due to the kinetic energy density of the particles and most of the volume of the gas was void of all matter. However, the known slow mixing of gases by diffusion revealed that, even though the mean speed of the particles was high, they could only travel a short distance, the mean free path, before colliding with other particles. This view of a gas as a chaotic region of constant motion and collisions helped other scientists to see the apparently homogeneous gas phase in a new light.

James Clerk Maxwell (1831-1879) was intrigued by the paper of Clausius. Even though he did not know whether there were such gas particles, he calculated the consequences of the theory (1860) (*3*). A system of colliding particles would transport momentum in the gas, and the gas viscosity would give a measure of the physical size of the particles. The collision diameter, σ, for typical gases was found to be fractions of a nanometer (*4*), and Maxwell gave a stirring lecture to the British Association in 1873 on the molecular view of matter. He distinguished his conclusions in terms of the completeness of his knowledge. He considered the relative masses of gas particles and the average velocities to be known with a high degree of precision. A lower degree of certainty was associated with the relative size of the gas particles and their mean free path. The lowest confidence was expressed with regard to the absolute mass, absolute size, and the number density of molecules in the gas. Josef Loschmidt (1821-1895) estimated the number density of the gas from a consideration of the measured diffusion and the concept of a collision diameter and a mean free path. He obtained a reasonable value for this quantity that is now called the Loschmidt number. Maxwell was then able to use gas viscosity data to infer the absolute size of the atoms.

If the particles were of finite size and attracted one another, the equation of state should reflect this fact. Precise measurements of the equation of state for carbon dioxide by Thomas Andrews (1813-1885) were reported to the Royal Society in 1869 as the Bakerian Lecture for that year (*5*). These measurements were explained by J.D. van der Waals (1837-1923) in terms of physical entities that were characterized by both a repulsive core and an attractive potential (*6*). This insight was summarized by the equation:

$$P(n,V,T) = \frac{nRT}{V-nb} - \left(\frac{na}{V}\right)^2 \qquad (1)$$

where n is the number of moles of gas particles, b is the occupied volume of a mole of particles and a is a measure of the attractive energy of the particles. The calculated size of the molecules was comparable to that inferred from the gas viscosity. But, what were these particles like?! They were not rigid balls, since they attracted one another at distances greater than the collision diameter.

Another observable property of gases is the heat capacity. The molar heat capacity of monatomic gases was measured and found to be equal to (3/2)R, the value predicted for a perfect (point particle) gas. But, actual atoms had a well defined physical size. Since finite spheres would be expected to rotate, where was the heat capacity due to rotation? Maxwell worried about this failure of the kinetic theory. Another type of eyes was required to see this result in its proper context.

Lord Rayleigh (1842-1919) probed the argon atoms he discovered with visible light. They scattered light in all directions. What was it that allowed the chemical atoms to interact with light? J.C. Maxwell explained light as an electromagnetic phenomenon. Were atoms electromagnetic objects as well? Rayleigh assumed that atoms were electrically polarizable and he had used Maxwell's equations to calculate the scattering of light (7). The success of this theory helped him to see his experiments as evidence for atoms.

Jean Perrin (1870-1942) pursued the nature of atoms his whole career (8). When the eyes of quantum mechanics were just developing in the 20th century, Perrin reasoned that the missing rotational heat capacity was due to a very small moment of inertia for atoms. This could only occur if the mass was concentrated in a very small volume. He "saw" atoms as spherical, but not as uniformly distributed masses. But, if most of the mass was concentrated in a very small volume, what determined the collision diameter measured from the gas viscosity?

The Eyes of Spectroscopy

The view of a chemical atom as a structureless hard sphere was shattered by the invention of the gas discharge tube. Michael Faraday (1791-1867) was able to produce high enough voltages to ionize gases confined to low pressure glass tubes. The phenomenology of these early discharge tubes was complicated, but they emitted light of both a continuous nature and a set of particular colors associated with the precise chemical species in the tube. Better tubes were produced by H. Geissler, and an atlas of chemical spectroscopic colors was rapidly obtained for many elements. How could "indivisible" atoms produce so many colors?

J.C. Maxwell saw these atoms as capable of internal vibrations (9). Atoms had some form of internal structure and were not static! If there were internal modes of motion in the atoms, Boltzmann predicted that equipartition of energy would lead to an increased heat capacity. Maxwell worried about the missing heat capacity due to these motions as well. Maxwell died before an explanation of the internal structure of atoms could be formulated, but he did have the eyes to see them as much more interesting than the ατομος of the Greeks.

The Eyes of Cathode Rays

Discharge tubes continued to be improved and Sir William Crookes (1832-1919) was able to obtain vacua low enough to allow another phenomenon to be observed: the wall of the tube glowed brightly when struck by "cathode rays" (10). His Bakerian Lecture at the Royal Society in 1879 was a masterful success. The extensive experiments of Crookes established that the new phenomenon was due to particles that could be deflected by a magnetic field and blocked by metal foils. They also heated the glass where they produced fluorescence. When a low pressure gas was subjected to a high voltage, Perrin (11) observed that the gas conducted electricity and that the cathode rays were the carriers of negative electricity. The carrier was discovered by Sir J.J. Thomson to be a very light particle with a charge equal to the unit of electricity (12). This "electron" could be removed from the atom and had properties that could be measured separately. Electrons obtained from different discharge tubes were all found to be identical. For this work Thomson received the Nobel Prize in Physics for 1906.

The discovery of the "atom of electricity" led to a flurry of speculation about the arrangement of the electrons inside the chemical atom. Intact chemical atoms are electrically neutral, but Perrin observed that when the cathode rays are produced, positive ions are also created. The notion that atoms are a combination of positive and negative parts is one of the key insights of this period. But the number of electrons and the nature of the positive part was still unknown.

The Eyes of Scattering

An even bigger surprise was soon observed by Henri Bequerel (1852-1908): some atoms needed no outside encouragement to emit massive particles with electric charges (13). The phenomenon of radioactivity changed the way we see atoms forever. Lord Ernest Rutherford (1871-1937) pursued the field and characterized the emitted particles; they were called alpha and beta particles. The alpha particles had a positive charge of two units and a mass equal to helium, and the beta particles were high energy electrons originating in the nucleus. For this work he received the Nobel Prize in Chemistry in 1908. One of the key steps in this successful experimental program was the development of a method to "see" the alpha particles. One approach used a gas discharge tube with a voltage very near the sparking point: the Geiger counter. When an alpha particle entered the vessel, the ionization of the gas was amplified and a spike was observed on the electrometer. The other approach used the phosphorescent properties of zinc sulphide; alpha particles produced visible scintillations when they struck a screen made of this mineral.

Lord Rutherford and his team of brilliant experimentalists proved beyond a reasonable doubt that alpha particles were doubly charged ions of helium (14). They also demonstrated spectroscopically that alpha particles became ordinary atoms of helium when they interacted with other atoms in a low pressure gas. What was not known was the internal structure of an ordinary atom of helium. For example, the number of electrons associated with a helium atom was unknown, and the size of the alpha particle was not yet known. Since alpha particles were emitted by the heavy radioactive elements, Rutherford speculated that rapidly moving helium atoms were a constituent of these atoms. It was supposed that the two electrons were "lost" on the way out of the atom.

Once the charged particles were obtained and characterized, they could be used as probes of intact atoms. The massive alpha particles obtained from radium were fired at thin metal targets and the scattered particles were detected as a function of scattering angle. Some of the particles were scattered through very large angles. In order to produce such an outcome, atoms must contain regions of mass and charge comparable in size and density to alpha particles. Rutherford produced a detailed theory of the scattering of charged particles by the massive nuclei of the heavier atoms (15). This theory was beautifully verified by the experiments of Hans Geiger and Sir Eric Marsden (16). It was also established experimentally that beta particles were scattered by the "nucleus" of the atom. The positive charge on the nucleus was measured to be approximately ½ the atomic mass of the heavy elements. The size of the nucleus could also be estimated from the scattering data and was found to be very much smaller than the overall atomic size. What Perrin inferred from thermodynamic data, Rutherford inferred from particle scattering: atoms contain a very dense nucleus.

With a growing understanding of the charge on the nucleus, the number of electrons needed to neutralize the charge was becoming clear. After the war, precision measurements of the scattering of alpha particles by Sir James Chadwick refined the value of the nuclear charge and demonstrated that it was equal to the atomic number multiplied by the magnitude of the charge on the electron (17).

Scattering experiments were also carried out on gaseous samples of hydrogen and helium. It was found that the nucleus of hydrogen contains a single positive charge and the nucleus of helium a doubly positive charge. The size of the nucleus was found to be in the femtometer range. Since the nucleus of helium has a mass approximately four times that of hydrogen, Rutherford speculated that it was composed of four hydrogen nuclei and two intranuclear electrons. There was something especially stable about this arrangement, since these helium nuclei were observed intact, as alpha particles, even after the trauma of radioactive emission. This theory of the constitution of atomic nuclei, expressed in the context of the triumphant paper of 1914 entitled "The Structure of the Atom" (18), was the "standard model" until the discovery of the neutron by Chadwick. Even though Rutherford speculated in 1920 that there might be neutral particles in the nucleus composed of a special state of a proton and an electron, the paradigm shift occurred only when actual neutrons were shown to exist and were characterized.

The Eyes of X-rays

When cathode rays impinge on metal targets, it is observed that a new kind of ray is produced: x-rays. The x-rays are very short wavelength electromagnetic radiation, comparable in size to atoms, and have proven to be ideal probes. X-rays are scattered by atoms, and J.J. Thomson produced a detailed theory that related the magnitude of this effect to the number of electrons in an atom (19). A value comparable to the atomic number was obtained for the light atoms in the gas phase. The result also depended on knowing the value of the Avogadro number, but by this time Perrin had established its value.

The existence of the Periodic Table of the chemical elements and the concept of the atomic number motivated A. van den Broek to assert that the nuclear charge in a neutral atom was exactly equal to its atomic number (20). This cogent speculation was given experimental support by the brilliant experimental work of H.G.J. Moseley (21). He measured the frequency, v , of the characteristic x-rays for most of the known elements using the recently discovered crystal monochromator. A relationship was established:

$$v = A(N - b)^2 \tag{2}$$

where N is the atomic number. At least two x-ray frequencies are observed for most elements and the value of A and b depended on whether the K or L x-rays were being observed. The remarkable precision of this simple formula for all the elements increased our confidence in the speculation of van den Broek.

The Eyes of Atomic Spectroscopy

As soon as the electron was discovered, speculations about the relationship between the line spectra of atoms and the motions of the electrons were offered. The experimental spectroscopists were also busy organizing their data into empirically pleasing forms. In 1890 J. R. Rydberg produced an equation that unified all the known data for hydrogen and the alkalis (22).

$$\frac{n}{N_0} = \frac{1}{(m_1 + \mu_1)^2} - \frac{1}{(m_2 + \mu_2)^2} \tag{3}$$

where n is the frequency in wavenumbers of the line, $N_0 = 109721.6$ is the Rydberg constant, the m's are integers and the mu's are empirical values associated with which series of lines is being considered. This suggested that the observed lines were due to energy differences between electronic levels in the atoms, but what were these levels?

That the observed lines were due to electrons was established by the experiments of Pieter Zeeman (1865-1943), who examined the sodium D lines when the atoms were in a magnetic field. The lines were broadened and the magnitude indicated that the effect was due to particles with the charge and mass of the electron (23). Another key observation was the polarization of the emitted light from different parts of the line. The light in the tails was circularly polarized, when observed along the direction of the magnetic field, while it was linearly polarized when observed at right angles to the field.

The key that unlocked the mystery of the line spectra was found with the simplest atom, hydrogen. It consists of a nucleus composed of a single positive charge and a single electron. Bohr made the assumption that since atoms are observed to be stable, recombination of an electron and a proton can only lead to certain discrete states (24). The energy of these states was calculated to be

$$W = \frac{2\pi m e^4}{h^2 \tau^2} \qquad (4)$$

where h is Planck's constant and τ is an integer. The atom can then emit light of a particular frequency by making transitions between these energy levels. The frequency of the light is given by:

$$v = \frac{2\pi^2 m e^4}{h^3}\left(\frac{1}{\tau_2^2} - \frac{1}{\tau_1^2}\right) \qquad (5)$$

The agreement between this relation and the observed line spectrum of hydrogen was far too good to be a mere coincidence. While the details of the electron orbits employed by Bohr in his calculation may not be a part of the current paradigm, the concept of electronic energy levels is here to stay.

The overall picture of the atom envisioned by Bohr was a dense nucleus of fixed charge surrounded by rings of electrons. The complicated optical spectra and the simple x-ray spectra suggested that the ring closest to the nucleus was different than the outer rings. More theory and more observations were necessary to refine this picture, but the shell theory of electronic structure has persisted.

The Eyes of Radioactive Elements

One of the most important observations of atoms is the set of relationships between elements that belong to one of the series of radioactive decays. The parent elements of uranium, thorium and actinium decay through many intermediates to the stable element lead. The Nobel Prize in Chemistry for 1921 was awarded in 1922 to Frederick Soddy for his complete characterization of these processes. The story is beautifully told in his Nobel Lecture entitled "The origins of the conception of isotopes" (25).

Uranium-238 emits an alpha particle to become an isotope of thorium. This unstable element emits a beta particle to become the element now known as Protactinium (Pa), which then emits another beta particle to become an isotope of uranium. This chain proceeds through another isotope of thorium, through radium, radon, polonium, bismuth, thallium and lead. The final product is lead-206. The series that starts with thorium-232 ends with lead-208. Soddy was able to isolate the different lead isotopes in high enough purity to demonstrate using chemical techniques that the atomic weights of two samples of lead with identical chemical and spectroscopic properties had different atomic weights. The final picture of these elements reveals that there are several isotopes for each of them.

Discussion of the composition of the nucleus was also furthered by the experimental discovery of the proton by Rutherford in 1918. While the singly positively charged nucleus of the hydrogen atom was known, it was not clear that nuclei of the higher elements contained such discrete particles. Rutherford directed alpha particles at nitrogen gas and found that particles with the mass and charge of the hydrogen nucleus were emitted. These particles were then identified as the proton. The picture of the nucleus that Soddy envisioned consisted of a number of protons equal to the mass number and enough electrons to yield a nett (common British usage of this period) charge equal to the atomic number. If an intranuclear electron was ejected, the atomic number then increased by one unit.

The Eyes of Mass Spectroscopy

Chemical techniques of analysis deal with a very large number of atoms and yield averages over the sample. Once the concept of isotopes was accepted, a search for different isotopes of every element was pursued. The key to the success of this search was the development of a precision instrument that sampled the atoms one at a time. It had been known since the development of the cathode ray tube that positive ions were also produced, and early experiments with these particles revealed singly and doubly charged species of the atoms and molecules that were contained in the tube. Sir J.J. Thomson observed in 1912 that when neon was the background gas, particles of mass number 20 and 22 were observed. Attempts to obtain pure samples of the two different atoms by fractionation techniques were unsuccessful, but in retrospect this was because they were both neon isotopes.

Francis William Aston (1877-1945) constructed a mass spectrograph using electromagnetic focusing that had a resolution better than 0.01 mass units. With this device he was able to demonstrate more than 200 naturally occurring isotopes of more than 30 elements. He received the Nobel Prize in Chemistry for 1922 and delivered a clear lecture on the measurement and meaning of isotopes (*26*). He noted that except for hydrogen, the measured atomic masses of each isotope were very near whole numbers. It appeared that the early speculations of Prout were returning to respectability. Aston envisioned the nucleus as "consisting of K+N protons and K electrons." This nuclear paradigm was good enough to allow the spectacular progress in the interpretation of the electronic spectra to occur. The current highly elaborated dynamic model of the nucleus could wait until another era.

The Eyes of Angular Momentum

While the theory of Bohr was a major step forward, and it helped to understand the observed hydrogen spectrum, it left many other observations in the dark. New light was shed on the subject of atomic structure and the line spectra by Arnold Sommerfeld (1868-1951) (*27*). He elaborated the basic theory of Bohr by observing that the orbits could also be elliptical, and that for each principal energy level, there could be a specific number of elliptical orbits of different angular momentum. This proliferation of orbits allowed many more electrons to be considered. Transitions between Sommerfeld orbits required a change in the angular momentum of the atom. Well in advance of the experimental proof that quanta of light, photons, have angular momentum, Sommerfeld concluded that they must, since angular momentum is conserved in the optical processes that lead to the line spectra. This insight opened the way for a whole generation of theorists, many of whom worked for Sommerfeld, to reason from observations to concepts that were either required by fundamental physics or were at least not forbidden. An especially good summary of the Sommerfeld approach was written by Leon Brillouin (28).

One important observation that demanded another great insight was the splitting of the yellow sodium D line into a doublet. Simple transitions between Sommerfeld orbits would yield single lines, in the absence of strong magnetic fields that were observed by Zeeman. G.E. Uhlenbeck and S. Goudsmit proposed that the effect was due to intrinsic angular momentum associated with the electrons themselves (29). The magnitude of this effect was calculated to be ½ that of the photon. The energy levels of the elliptical orbits in sodium could then be split into ones where the electron angular momentum either added or subtracted from the orbital angular momentum. Excellent agreement was obtained for the value of the splitting of the sodium D line.

Using the Bohr-Sommerfeld orbits and the concept of electron spin, Brillouin was able to explain the complicated line spectra of atoms and to calculate the term symbols for the atomic states. While the current paradigm is usually expressed in terms of the language of wave mechanics, Brillouin succeeded in the conceptual world of the previous paradigm. The electronic structure of the chemical elements could be rationalized in terms of the different orbits. A detailed treatment of the Zeeman effect gave the correct number of orbits associated with each magnitude of angular momentum. The spectacular success of the current view of the atom should not prevent us from seeing the more complicated situation during this exciting and tumultuous time in the history of the atom.

Concluding Thoughts

The picture of the chemical atom has changed substantially since the time of Dalton, but the interplay of bold thoughts and new observations has been an integral part of this story. The static atoms of the past have given way to the very dynamic atoms of the present as new particles and new ideas were discovered. The current picture of the chemical atom is grounded in the key observations presented above. Since actual chemical atoms are still very far from everyday experience, it remains a challenge to be able to see the atoms in the results of laboratory experiments. A recounting of the efforts of fellow chemists and physicists from the past to see the atoms encourages us to follow in their paths.

References

1. Thackray, A. *John Dalton: Critical Assessments of His Life and Science*; Harvard University Press: Cambridge, 1972.
2. Thomson, J. J. *The Atomic Theory, The Romanes Lecture (1914)*; Oxford/ Clarendon Press: Gloucestershire, U.K., 1914.
3. Maxwell, J. C. *Philos. Mag.* **1860**, *4*, 19.
4. Maxwell, J. C. *Nature* **1873**, *8*, 437.
5. Andrews, T. *Philos. Trans. R. Soc.* **1869**, *159*, 575.
6. Van der Waals, J. D. *On the Continuity of the Gaseous and Liquid States*; A. W. Sijthoff: Leiden, The Netherlands, 1873.
7. Strutt, J. W. *Philos. Mag.* **1871**, *41*, 107, 274, 447.
8. Perrin, J. *Les Atomes*; Felix Alcan: Paris, 1913.
9. Maxwell, J.C. *Theory of Heat*; Longmans, Green and Co.: London, 1871.
10. Crookes, W. *Philos. Trans. R. Soc.* **1879**, 169, 135.
11. Perrin, J. *Compt. Rend.* **1895**, *121*, 1130.
12. Thomson, J. J. *Philos. Mag.* **1897**, *44*, 293.
13. Becquerel, H. *Compt. Rend.* **1896**, *122*, 420.
14. Rutherford, E.; Soddy, F. *Philos. Mag.* **1903**, *5*, 453.
15. Rutherford, E. *Philos. Mag.* **1911**, *21*, 669.
16. Geiger, H.; Marsden, E. *Philos. Mag.* **1913**, *25*, 604.
17. Chadwick, J. *Philos. Mag.* **1920**, *40*, 734.
18. Rutherford, E. *Philos. Mag.* **1914**, *27*, 488.
19. Thomson, J. J. *Philos. Mag.* **1906**, *11*, 769.
20. Van den Broek, A. *Philos. Mag.* **1914**, *27*, 455.
21. Moseley, H. G. J. *Philos. Mag.* **1913**, *26*, 1024;**1914**, *27*, 703.
22. Rydberg, J. R. *Philos. Mag.* **1890**, *29*, 331.
23. Zeeman, P. *Nature* **1897**, *55*, 347.
24. Bohr, N. *Philos. Mag.* **1913**, *26*, 1.
25. Soddy, F. In *Nobel Lectures*; Elsevier: Amsterdam, 1966.
26. Aston, F. W. In *Nobel Lectures*; Elsevier: Amsterdam, 1966.
27. Sommerfeld, A. *Atomic Structure and Spectral Lines*; Methuen: London, 1923.
28. Brillouin, L. *L'Atome de Bohr*; Presses Universitaires de France: Paris, 1931.
29. Uhlenbeck, G. E.; Goudsmit, S. *Nature* **1926**, *117*, 264.

Chapter 8

Rediscovering Atoms: An Atomic Travelogue

A Selection of Photos from Sites Important in the History of Atoms

Jim Marshall* and Jenny Marshall

University of North Texas, Denton, TX 76207
*jimm@unt.edu

This chapter outlines visits to several sites where important discoveries in the history of the atom took place. Connecting with artifacts and locations associated with specific historical episodes can make those developments appear more salient.

For many years, we have been following the footsteps of the discoverers of chemical elements. We have traveled extensively to places associated with the various elements—to the sites of mines, laboratories, museums and other locations where work on the discovery of elements was carried out or where artifacts are displayed and interpreted. Under the title "Rediscovery of the Elements" (*1*), we have compiled guides to these sites to allow students, educators, and other curious people to follow along, whether actually or vicariously. The project includes extensive photographs as well as directions and coordinates.

This chapter draws upon materials gathered on our travels relevant to the history of atoms. Much of the research described in this chapter is explained in greater detail elsewhere in this book. This chapter selects a few places where one can make contact with fundamental discoveries about the atom and the people who made them.

Manchester, England — Dalton

John Dalton, regarded by most chemists as the originator of the first scientifically fruitful chemical atomic theory, lived and worked in Manchester, England, for much of his life. Several commemorations of Dalton can be found in Manchester, from an unobstrusive plaque on the site where Dalton's laboratory once stood to a bronze statue outside the John Dalton building of Manchester Metropolitan University.

One of the visually most interesting commemorations of Dalton in Manchester is a painting in the Great Hall of the Manchester Town Hall. "Dalton Collecting Marsh-Fire Gas," painted by Ford Maddox Brown, is shown in Figure 1. Dalton appears to have the assistance and attention of local children in this activity.

The site of the Dalton plaque had a rich history. Dalton's laboratory was in the premises of the Manchester Literary and Philosophical Society, a learned society founded in 1781. He was a member of the Society from 1794 until his death in 1844, serving as President for 28 years. The building at 36 George Street was built by the Society in 1799 and was its headquarters until a bombing raid destroyed it in 1940. Many of Dalton's papers were destroyed along with the building. Some of his possessions survive at the Manchester Museum of Science and Industry.

Figure 1. Painting by Ford Maddox Brown, "Dalton Collecting Marsh-Fire Gas," in Manchester Town Hall. (Photo Copyright J. L. and V. R. Marshall.)

Crown and Anchor, London, England — Davy, Wollaston, Thomson

The British Royal Society held dinners at the Crown and Anchor tavern from the late 18th century through the middle of the 19th century. Much scientific discussion occurred in that building on the Strand, opposite the church of St. Clements. The tavern is the frontmost of the block of buildings shown at right in Figure 2. Today, the church of St. Clements remains, but office buildings have taken the place of the tavern.

Humphry Davy, William Hyde Wollaston, and Thomas Thomson were among the prominent chemistry Fellows of the Royal Society during this time. In 1807 and 1808, they were discussing multiple proportions and the logic of the atomic hypothesis. Thomson's 1807 *A System of Chemistry* (*2*) presented aspects of Dalton's atomic theory (with permission) the year before the publication of his own *New System of Chemical Philosophy*. Thomson presented to the Royal Society work on combining ratios in salts of oxalic acid (salts we would identify as oxalates and binoxalates) (*3*). Wollaston followed this paper with one on carbonates and bicarbonates (*4*). He regarded his results as examples of Dalton's general observation that compounds form in simple ratios of atoms. Thomson, founder and editor of the journal *Annals of Philosophy*, was an early advocate of Dalton's atomic theory.

Figure 2. Environs of church of St. Clements, London, including the Crown and Anchor tavern, lower right.

le Societé d'Arcueil, Arcueil, France — Berthollet and Gay-Lussac

Research by the French natural philosophers Joseph-Louis Gay-Lussac and Pierre-Louis Dulong during the early 19th century supported the new atomic theory. That work was done in the laboratory of Claude-Louis Berthollet, the founder of the Société d'Arcueil, near Paris. Berthollet's home is shown in Figure 3. The site of the home today is marked by a plaque, shown in Figure 4. A bust of Berthollet can be found in Arcueil's city hall, the Centre Marius Sidobore. Arcueil itself lies just south of the Boulevard Périphérique that rings Paris.

Berthollet's property might seem an unusual stop on a tour of atomism, given that he did not believe in atoms. His analytical work made him skeptical of the law of definite proportions that emerged around the turn of the 19th century. Berthollet, on the contrary, found examples of variable proportions. The notion of compounds arising from the union of definite small numbers of atoms, which was a logical explanation of definite proportions, was difficult to reconcile with variable proportions.

Joseph-Louis Gay-Lussac's memoir on the combining volumes of gases (5) contained data that Amedeo Avogadro would soon interpret atomistically (6). Gay-Lussac was a protégé and assistant of Berthollet, and he presented this memoir before the Société d'Arcueil. What Gay-Lussac reported is that many reactions of gases occur in ratios of small whole numbers by volume, such as two of hydrogen to one of oxygen to form water. Avogadro noted that if equal volumes of gases contained equal numbers of atoms or molecules, then the reactions themselves involved small whole-number ratios of atoms—just as Dalton had proposed.

Neither Gay-Lussac nor Berthollet accepted this atomistic interpretation, though. To be sure, there was a significant stumbling block: how could two atoms of hydrogen combine with one of oxygen to yield **two** atoms of water? That is, how could the "atom" of oxygen be split in the course of this reaction? Avogadro had an answer to this objection, essentially the answer that we give today, distinguishing between atoms and molecules and positing that hydrogen and oxygen were diatomic molecules. But Avogadro had no direct or independent evidence for this explanation, which also contradicted notions of chemical affinity prevalent at the time.

Dulong was also an associate of Berthollet and a member of the Société d'Arcueil. His 1819 paper on heat capacities of elements in collaboration with Alexis-Thérèse Petit was widely interpreted as support for the atomic hypothesis. They noted that the product of specific heat times atomic weight was very nearly the same for a large number of solid elements. They recognized that the quantity in question represents the heat capacities of the atoms—or in modern terms, molar heat capacities. And they generalized the results, asserting that, "atoms of all simple bodies have exactly the same capacity for heat." (7)

Figure 3. The house of Claude-Louis Berthollet, founder of le Société d'Arcueil near Paris.

Figure 4. Plaque marking the site of Berthollet's home. Translation: "Claude Berthollet (1748-1822) lived on this property. Founder of industrial chemistry, he established the Arcueil Society of Chemistry and Physics in 1807. He was mayor of this town in 1820. Gift of the people of Arcueil". (Photo Copyright J. L. and V. R. Marshall.)

Heidelberg, Germany — Bunsen and Kirchhoff

Heidelberg, Germany, contains many memorials and artifacts of Robert Bunsen and Gustav Kirchhoff, the inventors of spectral analysis.

They introduced their spectroscope in a paper published in 1860 (8). They emphasized the utility of the spectroscope as a very sensitive tool for qualitative elemental analysis. They predicted that the tool would be valuable in the discovery of yet unknown elements. They noted that the spectroscope had convinced them of the existence of another alkali metal besides lithium, sodium, and potassium; eventually they found two—cesium and rubidium. In that 1860 paper, they noted that their instrument could shed light on the chemical composition of the sun and stars—not many years after Auguste Comte wrote that such knowledge was beyond the reach of human beings.

Figure 5 shows the spectroscope as depicted in their paper (above) and on display at Heidelberg University (below). Note the flame source—a Bunsen burner, of course. The display is in the chemistry department at the University's new campus in Neuenheim, across the river from the old city.

Kirchhoff would distinguish three kinds of spectra: continuous spectra of black-body radiation (a term he coined), bright-line spectra from hot sources, and dark-line spectra of light passing through cool samples. Already by 1860 he recognized that the bright-line emission spectra of hot gases are coincident with the dark-line absorption spectra of cool gases.

Spectroscopy was to prove indispensable in unlocking the structure of atoms, particulary their electronic structure—but those developments would depend on other, later researchers. Max Planck's analysis of blackbody radiation and Bohr's theory of the hydrogen spectrum are just two examples.

The old city is where Bunsen and Kirchhoff worked. Figure 6 shows a statue of Bunsen on the main street of the old city of Heidelberg. The statue is in front of a building where Kirchhoff lived. Across the street is the building, shown in Figure 7, where Kirchhoff developed a theory and method of spectroscopy and where he and Bunsen discovered cesium and rubidium. The building, "Zum Reisen" had been a distillery in the 18th century before the University had acquired it. The unassuming plaque on the building says (in translation), "In this building in 1859, Kirchhoff founded spectral analysis with Bunsen and applied it to the sun and stars, thereby opening the study of the chemistry of the universe."

Bunsen was quite imposing physically. He was tall (six feet) and he has been described as "built like Hercules." (9)He was apparently impervious to pain, for he was said to be able to handle hot objects with total disregard, picking up the lid of a glowing porcelain crucible with his bare fingers (10). When he was blowing glass, one could sometimes smell burnt flesh, according to English chemist Henry Enfield Roscoe, who worked with Bunsen in Heidelberg (11). A modest man of simple manners, Bunsen placed great value on facts and little on theories or systems. In his last lectures in 1889, Bunsen did not refer to the periodic law, despite the fact that both of its principal formulators, Dmitri Mendeleev and Julius Lothar Meyer, had worked with him in Heidelberg.

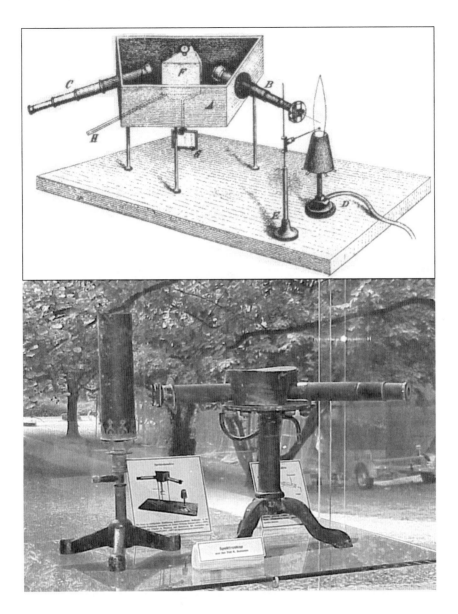

Figure 5. The spectroscope of Bunsen and Kirchhoff. Above, figure from reference (8). Below, photo of original spectroscope on display case at Heidelberg University. (Photo Copyright J. L. and V. R. Marshall.)

Figure 6. Statue of Robert Bunsen on the main street of Heidelberg, Germany. (Photo Copyright J. L. and V. R. Marshall.)

Figure 7. "Zum Reisen," the former distillery where Kirchhoff and Bunsen invented spectral analysis. (Photo Copyright J. L. and V. R. Marshall.)

In Germany with Julius Lothar Meyer

While in Germany, one can visit several sites from the life and career of Meyer. He shared the 1882 Davy Medal of the Royal Society (London) with Mendeleev for discovery of the periodic law. Today, Mendeleev is the first name associated with the discovery of the periodic law and invention of the periodic table. In most accounts, though, Meyer stands second.

Meyer included a partial periodic table of 28 elements in the first edition of his *Modernen Theorien der Chemie* published in 1864. The table only included slightly more than half of the elements then known, but those elements are arranged in order of increasing atomic weights and aligned in columns according to valence. While preparing a second edition of the book in 1868, Meyer prepared a more comprehensive table, which he did not publish (*12*). He did publish a periodic table in 1870 in *Annalen* (*13*). That paper included a plot of atomic volume that displays the periodicity of that elemental property as well as a periodic table that many consider superior to Mendeleev's 1869 table.

Meyer was a professor at the Forstakademie (Forestry School) in Eberswalde when he formulated his unpublished comprehensive table. The building where he worked is shown in Figure 8. (Eberswalde is in the northeast of Germany, northeast of Berlin, not far from the Polish border.) Meyer moved to the Karlsruhe Polytechnikum in 1868, and he left his table in Eberswalde with his successor, Adolf Remelé. Carl Seubert, one of Remelé's colleagues, published that table in 1895 after Meyer's death (*12*).

Figure 8. Old Forest Academy building in Eberswalde, Germany, where Julius Lothar Meyer drafted his first comprehensive periodic table. (Photo Copyright J. L. and V. R. Marshall.)

Figure 9. Columns in Varel, Germany, bearing sculpted heads of Julius Lothar Meyer, Dmitri Mendeleev, and Stanislao Cannizzaro. (Photo Copyright J. L. and V. R. Marshall.)

Meyer was born in 1830 in Varel, not far from the North Sea in what is now Germany. (At the time of Meyer's birth it had been part of the Duchy of Oldenburg.) His birthplace is marked by a plaque, and there is a school named for him, Lothar-Meyer-Gymnasium. A more interesting memorial is shown in Figure 9, three columns bearing sculpted heads of Meyer, Mendeleev, and the Italian chemist Stanislao Cannizzaro.

In 1860, these three chemists were all together in the flesh elsewhere in Germany. They were all among the attendees of the first international congress of chemists held that year in Karlsruhe. The purpose for gathering chemists from throughout Europe was to discuss and if possible define such important chemical terms as atom, molecule, and equivalent. Although the attendees were mindful that they had no authority to legislate on such matters, they hoped to bring clarity to the questions they would discuss.

In retrospect, the Karlsruhe Congress brought about widespread agreement on a system of atomic weights, and Cannizzaro deserves much of the credit for it. He spoke in the conference hall on reliable methods for determining atomic weights based on Avogadro's hypothesis, vapor densities, and specific heats. He also distributed a reprint of his sketch of a course of chemical philosophy, published two years earlier (*14*). Meyer later recalled reading Cannizzaro's pamphlet on his way home from the conference: "It was as though the scales fell from my eyes." (*12*) Historians of the periodic law consider the development of a consistent set of atomic weights to have been a prerequisite to the discovery of the periodic law and the Karlsruhe Congress a key event.

Figure 10. Ständehaus (right) in Karlsruhe, Germany in 1860 (above) and at present (below). (Photo (below) Copyright J. L. and V. R. Marshall.)

The Congress met in the Ständehaus, the home of the parliament of the Grand Duchy of Baden, courtesy of the Archduke. That building is no longer in existence; however, its modern replacement evokes the style of the old one. (Figure 10 shows exterior views of the old Ständehaus (above) and the new one (below).) The new Ständehaus contains photos, displays, and other records of the original.

While in Karlsruhe, one can visit the building where Meyer worked at the Polytechnicum (now part of Karlsruhe Universität), but there are no memorials to him there. Karlsruhe is one of three cities in southwestern Germany where Meyer lived and worked. As mentioned above, he worked with Bunsen in Heidelberg. Tübingen is the third city. Meyer spent the last 20 years of his life as professor at its university, and he died there in 1895. The university now has a geology building named in his honor.

McGill University, Montreal, Canada — Rutherford and Soddy

The last sites visited in this chapter are associated with Ernest Rutherford and Frederick Soddy, pioneers in the study of radioactivity. Radioactivity is one of the phenomena that led chemists and physicists to understand that atoms were not indestructible or indivisible.

Rutherford spent about a decade in the Macdonald Chair of Physics at McGill University in Montreal, Canada. The old physics building, where he worked, is shown in Figure 11. He arrived there originally from New Zealand by way of Cambridge, England, where he had worked with J. J. Thompson. While at McGill, Rutherford discovered a radioactive "emanation" from thorium, which we know as radon (*15*). He characterized the time-dependence of radioactive emission (exponential decay) and applied the term half-life to the phenomenon (*16*). He studied the α particle extensively, beginning the series of experiments that would lead to the discovery of the nucleus (*17*). And, working with Frederick Soddy, he carried out the research for which he would receive the 1908 Nobel Prize in Chemistry. Soddy arrived at McGill from Oxford in 1900 to take the post of Demonstrator of chemistry. His was there for only about two years before returning to England to work with Sir William Ramsay at University College, London.

Figure 11. Old Physics Building at McGill University, Montreal. Here Ernest Rutherford and Frederick Soddy discovered the chemical transformations that accompany radioactive emissions. (Photo Copyright J. L. and V. R. Marshall.)

104

Rutherford and Soddy recognized in 1902 that chemical transformations accompanied the emission of radioactive particles. The chemical transformations were far from obvious. The readily observable phenomenon in radioactivity is the penetrating radiation; the material that emits the radiation appeared, to no less an observer than Marie Curie, to be unchanged. Rutherford and Soddy saw the change, and they inferred (partly on the basis of Curie's work) that these chemical changes were at the atomic or subatomic level. "Since, therefore, radioactivity is at once an atomic phenomenon and accompanied by chemical changes in which new types of matter are produced, these changes must be occurring within the atom, and the radioactive elements must be undergoing spontaneous transformation. ... Radioactivity may therefore be considered as a manifestation of subatomic chemical change." (*18*) More than 100 years later, it is difficult to realize just how radical this assertion was: atoms were falling apart on their own, in the process changing into atoms of other elements.

At McGill, there is a display on Rutherford including a sculpted bust, descriptions and diagrams of his research, and some pieces of experimental apparatus. There is also a plaque in honor of Soddy at McGill, noting the time he spent as a "member of the staff in the chemistry department" and describing his collaboration with Rutherford. Their investigations "led to discoveries of fundamendtal importance," according to the plaque, "including the natural transmutation of elements." By the time the plaque was erected, the research was not only accepted but acclaimed; so the plaque could use the word that Rutherford and Soddy dared not—transmutation.

Glasgow, Scotland — Soddy

Rutherford was the sole recipient of the 1908 Nobel Prize in chemistry, "for his investigations into the disintegration of the elements, and the chemistry of radioactive substances." By that time, he was back in England, in Manchester, where we began this chapter.

Soddy did not share that Nobel, but he would win one of his own in 1921 "for his contributions to our knowledge of the chemistry of radioactive substances, and his investigations into the origin and nature of isotopes." The research conducted at McGill with Rutherford certainly falls under the first part of the citation.

The latter part of the citation covers work Soddy carried out while at the University of Glasgow. While there, he developed the displacement law of radioactive transformation, whereby an emitter of α radiation is displaced two places to the left in the periodic table (*i.e.*, it is transformed into the element two to the right) and an emitter of β radiation is displaced one to the right. He also introduced the term isotope, a word suggested to him in the building shown in Figure 12.

Figure 12. George Service House, University Gardens, Glasgow, Scotland, where the term isotope was born. (Photo Copyright J. L. and V. R. Marshall.)

Now we describe isotopes as atoms of the same element that have different nuclei, usually because they have different number of neutrons in the nucleus. Soddy introduced the term to describe elements that "occupy the same place in the periodic table." He said that "isotopes" or "isotopic elements" were chemically identical and, in most respects that do not depend on the atomic mass, physically identical as well (*19*). The term is derived from the Greek iso- (same) and topos (place).

Soddy introduced the term, but it was Dr. Margaret Todd who coined it. This was done at a dinner party in 1913 at the home of Soddy's father-in-law, Sir George Beilby. A plaque on the house marks the occasion.

References

1. Marshall, J. L.; Marshall, V. R. Rediscovery of the Elements, 2009. http://www.jennymarshall.com/rediscovery1.htm.
2. Thomson, T. *A System of Chemistry*, 3rd ed.; Bell & Bradfute; E. Balfour: London, 1807; Vol. 3.
3. Thomson, T. On oxalic acid. *Philos. Trans. R. Soc. London* **1808**, *98*, 63–95.
4. Wollaston, W. H. On super-acid and sub-acid salts. *Philos. Trans. R. Soc. London* **1808**, *98*, 96–102.
5. Gay-Lussac, J. L. *Mémoires de la Société d'Arcueil* 1809; Vol. 2. English translation in *Foundations of the Molecular Theory*, Alembic Club Reprints #4; Alembic Club: Edinburgh, U.K., 1911.

6. Avogadro, A. *J. Phys. (Paris)* **1811**, *73*, 58–76. English translation in *Foundations of the Molecular Theory*, Alembic Club Reprints #4; Alembic Club: Edinburgh, U.K., 1911.

7. Petit, A. T; Dulong, P. L. Recherches sur quelques points importants de la théorie de la chaleur. *Ann. Chim. Phys.* **1819**, *10*, 395–413. English translation in *Ann. Philos.* **1819**, *14*, 189–198.

8. Kirchhoff, G.; Bunsen, R. *Ann. Phys. Chem. (Poggendorff)* **1860**, *110*, 161–189.

9. McCay, L. W. My student days in Germany. *J. Chem. Educ.* **1930**, *7*, 1081–1099.

10. Partington, J. R. *A History of Chemistry*; Macmillan: London, 1961; Vol. 4.

11. Roscoe, H. E. *The Life and Experiences of Sir Henry Enfield Roscoe, D.C.L, LL.D., F.R.S.*; Macmillan: London, 1906.

12. Scerri, E. R. *The Periodic Table: Its Story and Its Significance*; Oxford University Press: Oxford, U.K., 2007.

13. Meyer, J. L. *Ann. Chem. Pharm.* **1870**, *7*, 354–364.

14. Cannizzaro, S. Sunto di un corso di filosofia chimica. *Nuovo Cimento* **1858**, *7*, 321–366. English translation as *Sketch of a Course of Chemical Philosophy*, Alembic Club Reprints #18; Alembic Club: Edinburgh, U.K., 1911.

15. Marshall, J. L.; Marshall, V. R. Ernest Rutherford, the "true discoverer" of radon. *Bull. Hist. Chem.* **2003**, *28*, 76–83.

16. Rutherford, E. A radioactive substance emitted from thorium compounds. *Philos. Mag.* **1900**, *49*, 1–14.

17. Giunta, C. G. Atoms are Divisible: The Pieces Have Pieces. In *Atoms in Chemistry: From Dalton's Predecessors to Complex Atoms and Beyond*; Giunta, C. G., Ed.; ACS Symposium Series 1044; American Chemical Society: Washington, DC, 2010; Chapter 6.

18. Rutherford, E.; Soddy, F. The cause and nature of radioactivity. *Philos. Mag.* **1902**, *4*, 370–396.

19. Soddy, F. Intra-atomic charge. *Nature* **1913**, *92*, 399–400.

Indexes

Author Index

Subject Index

D

Dallas, Duncan, 61
Dalton, John, 94, 94*f*
  A New System of Chemical Philosophy,
    1, 95
Daltonian atoms and molecules, 16*f*
Davy, Humphry, 95, 95*f*
de Morveau, Guyton, 12
De Rerum Natura, 24, 26
Democritus, 8, 24
Descartes, René, 10, 27
Duhem, Pierre, 63
Dulong, Pierre-Louis, 96
Dumas, Jean-Baptiste André, 37
Displacement reactions, 12*f*

E

Electron cloud, 19*f*
Electrons, 66
Empedocles, 23
Epicurus of Samos, 8, 24
Erlenmeyer, Emil, 45

F

Falconer, Isobel, 68
Faraday, Michael, 85
Foster, George Carey, 61
Frankland, Edward, 36, 60
Freind, John, 11

G

Galen of Pergamum, 24
Galilei, Galileo, 27
Gas pressure, 17*f*
Gassendi, Pierre, 9, 27
Gaudin, Marc Antoine Augustin, 36
Gay-Lussac, Joseph-Louis, 96
Geiger, Hans, 71, 72*f*, 73, 73*f*, 76*f*
Geissler, H., 85
Gerhardt, Charles Frédéric, 37
Gors, Britta, 61

H

Heat capacity, 85

Heraclitus, 23
Higgins, Bryan, 28
Higgins, William, 28

J

Joliot-Curie, Frédéric and Irene, 76

K

Kalàm, 25
Kaufmann, Walter, 67
Keill, John, 11
Kekulé, Friedrich August, 38, 40*f*, 61
Kinetic theory, 84
Kirchoff, Gustav, 98, 99*f*, 100*f*
Klason, J. P., 67
Klein, Ursula, 61
Kolbe, Hermann, 48
Kragh, Helge, 68

L

Laurent, Auguste, 36
Laertius, Diogenes, 8
Larmor, Joseph, 66
Lavoisier, Antoine-Laurent, 28
Le Bel, Joseph Achille, 50*f*, 51
Lemery, Nicholas, 27
Lenard, Philipp, 68
Leucippus, 8, 24
Liebig, Justus von, 37
Lives of Eminent Philosophers, 8
Lomonosov, Mikhail Vasilyevich, 28
Lorentz, Hendrik, 66
Loschmidt, Johan Josef, 47, 49*f*, 84

M

Mach, Ernst, 63
Marsden, Eric, 87
Marsden, Ernest, 73, 73*f*, 76*f*
Mass spectroscopy, 90
Maxwell, James Clerk, 66, 84, 85
Mayer, Alfred, 68
Mayow, John, 10
Mendeleev, Dmitri, 102
Meyer, Julius Lothar, 17, 101, 101*f*, 102*f*,
    103*f*